# Distillation Symposium 2018



Topical Conference at the 2018 AIChE Spring Meeting and 14th Global Congress on Process Safety

Orlando, Florida, USA
22 - 25 April 2018

ISBN: 978-1-5108-6430-6

**Printed from e-media with permission by:**

Curran Associates, Inc.
57 Morehouse Lane
Red Hook, NY 12571

**Some format issues inherent in the e-media version may also appear in this print version.**



Printed by Curran Associates, Inc. (2018)

For permission requests, please contact AIChE
at the address below.

AIChE
120 Wall Street, FL 23
New York, NY 10005-4020

Phone: (800) 242-4363
Fax: (203) 775-5177

www.aiche.org

**Additional copies of this publication are available from:**

Curran Associates, Inc.
57 Morehouse Lane
Red Hook, NY 12571 USA
Phone: 845-758-0400
Fax: 845-758-2633
Email: curran@proceedings.com
Web: www.proceedings.com

# TABLE OF CONTENTS

**Some abstracts in this book refer to images that were not made available by the publisher**

# 513345 Distillation Column Dynamics

**Abstract:**

Dynamic process simulations are increasingly popular for modeling the transient responses of distillation columns. The effectiveness of these simulations greatly depends on the models used to represent the hydraulics of a distillation column. In this paper, using transient data from FRIâ€™s distillation columns, models for estimating the dynamics of distillation columns will be examined and evaluated.

# 513662 A New Pilot Scale Facility for Testing High Pressure Cryogenic Distillation in LNG Applications

**Abstract:**

Liquefied Natural Gas (LNG) processes require cryogenic distillation columns to separate heavy hydrocarbon compounds from natural gas by operating at high pressure (4 to 6 MPa), over a wide range of temperatures (-60°C to 25°C ). Prediction of the column distillation performance requires both accurate vapour liquid equilibrium data (VLE) models and overall column efficiencies as a function of flow rate. Kister [1][2] has shown that errors in VLE data can have a major effect on mass transfer rates and tray efficiency calculations. An improved ability to describe this critical separation process will reduce operational costs and downstream blockage risks in mega-scale LNG plants, as well as lead to more efficient column design in small scale and/or floating LNG applications.

The Commonwealth Science and Research Organisation (CSIRO) in partnership with the University of Western Australia (UWA) have developed a unique pilot-scale cryogenic distillation column (CDC). The columns dimensions are 2m high and 50 mm internal diameter. It can operate at flow rates from 3 SLPM to 20 SLPM at a pressure of up to 50 bar. The column temperature is controlled with 3 independent systems that allows the overhead product stream to be set at temperature down to 60°C while the bottom stream is produced at up to 25°C. Full compositional analyses are performed on-line for both product streams and feed. Control schemes for column pressure and product flow rates are implemented on the overhead and bottom streams, respectively. Initial tests have been performed with Pyrex bead packing; Oldershaw trays will be installed in the near future. Fair et al. [3] have shown that Oldershaw column types are well suited for upscaling results obtained in the laboratory to industrial scale applications.

Initial tests have demonstrated successful control of the column temperature, flow and pressure as well as effective separation of a multi-component natural gas mixture spiked with additional LPG components to better represent industrial feed compositions. Steady state and dynamic process simulations of the column have been constructed to allow comparison of rate-based and equilibrium models with the measured data. This work will present the results of parametric studies with the column where feed location, flow rate, temperature and pressure are varied and discrepancies with existing models identified. Future improvements to the column will be discussed, including the implementation of a reflux condenser. Future work will also include tests of heavy hydrocarbon (BTEX) carry-over.

[1] Kister HZ. Effects of design on tray efficiency in commercial towers. Chem Eng Prog 2008;104:39–47.

[2] Kister HZ, Mathias P, Steinmeyer DE, Penney WR, Crocker BB, Fair JR. Equipment for Distillation, Gas Absorption, Phase Dispersion, and Phase Separation. Perry's Chem. Eng. Handbook, Eighth Ed., 2008, p. 14–50.

[3] Fair JR, Null HR, Bolles WL. Scale-up of Plate Efficiency from Laboratory Oldershaw Data. Ind Eng Chem Process Des Dev 1983;22:53–8.

## 510624 *Process Equipment Imaging By Tomographic Gamma Scan*

**Abstract:**

*Gamma scanning is a nuclear inspection technique widely used for troubleshooting industrial process equipment in refineries and petrochemicals. A radiation source and detector move vertically along the height of the vessel and the radiation readings are used to plot the graph of density profile. In recent years, many improvements have been added to this technique, but essentially the result of gamma scanning still consists in a 1-D density plot. We presented, at the 2016 Spring Meeting, the tomographic gamma scan that uses image reconstruction techniques to show the result of gamma scan as a 2-D density distribution image. In this work, we present the results obtained applying the innovative technique for the first time in a real operating process equipment. The obtained images match well with the 1-D density plots obtained by the conventional gamma scan. However, they also reveal a possible process problem that has never been visualized before with the traditional technique. The tomographic gamma scan can be immediately applied in practice with low financial investment.*

# 512171 Debottlenecking of a Large Industrial Deethanizer By Gravity Driven Two-Phase Flow Analysis

**Abstract:**

A large industrial Deethanizer in a NGL fractionation plant located in Mont Belvieu, Texas USA exhibited performance issues in its dehydration section since start-up. This problem forced the operations team to reduce rates and to operate at higher than design pressure to avoid hydrates. Troubleshooting efforts to resolve the problem included mechanical and hydraulic analysis of the internals and performing several column gamma scans. The mechanical integrity of the internals was suspected to be part of the problem. However, several gamma scans on the tower showed flooding conditions in the top of the dehydration section. A fluid dynamics analysis of the hydrocarbon drain pipe bypassing the dehydration section indicated the presence of two-phase flow.

The two-phase flow in the drain pipe explained the flooding conditions at the top of the dehydration section. A gamma scan of the drain pipe was performed and confirmed specific gravity differences at different elevations. The implementation of simple but unique piping modifications addressed the multiphase flow in the hydrocarbon line. The Deethanizer is currently operating at normal production rates.

Today, practicing engineers are still unable to predict with confidence whether a two-phase flow will develop. Many similar piping configurations are found in the chemical process industry, and this paper attempts to provide insights to prevent a recurring design problem.

# 512616 Loss of Float Valves Causing Flood

**Abstract:**

**Can Loss of Valve Floats Lead to Premature Flood?**

**By Matthew Olsson and Henry Kister, Fluor**

An Atmospheric Crude Tower has functioned well since 1989. All the reflux to the top of the tower was supplied by a top pumparound (TPA) loop. In an attempt to lower the naphtha 95% point, the TPA was supplemented by injecting cold reflux from the overhead receiver. As this reflux is believed to contain water, a filter and coalesce were added to remove this water prior to tower entry. However, frequent plugging of the filter necessitated bypassing, causing corrosive water to enter. In the tower it hydrolysed chlorides, forming HCl, that caused extensive corrosion and damage to the TPA trays, resulting in flooding and high naphtha end points.

Pressure drop measurements showed flooding taking place both in the 4-pass TPA trays and also in the 2-pass fractionation section below. A gamma scan confirmed flooding in the upper four-five trays in the fractionation section. Upon shutdown and inspection, extensive corrosion and damage were observed in the TPA trays, as expected. The surprise was that the upper trays in the fractionation section were found clean, but missing most of the valve floats. The missing valve floats were not found anywhere, probably washed away or corroded away. This increased the open area on the trays and should have moved the trays further away from flood, but the flood only occurred near the end of that run, and did not occur in the earlier runs, when the valve floats are believed to have been in place.

This raised the question: could losing the valve floats have promoted flood in the fractionation trays? In an endeavour to gain better understanding to the observation, we applied the Fluor Multipass Maldistribution Model (MMM) to see whether these observations can be modelled. In a previous article[1], we applied the same model to discover multiple steady state vapor/liquid distributions in 2-pass moving valve trays at turndown.

Due to the symmetry of 2-pass trays, a perfect split of both vapor and liquid between the passes is always one possible and well-known steady state distribution. Our analysis shows that uneven or complete float removal on the trays can channel the vapor and liquid in different directions, which may account for the premature flood and above observations.

The trays were repaired and the tower restarted. To prevent recurrence, the refinery stopped supplementing the TPA reflux by injecting cold reflux, even for short periods, and guidelines were drawn for the maximum moisture content in the reflux. The tower has been performing well since.

This experience teaches how damaging the cold reflux can be to column internals in the top sections. On a different topic; it teaches that losing valve floats can lead to channelling and premature flood.

1. Olsson, M., and H. Z. Kister, Multiple Steady States in 2-Pass Hydraulics of Moving Valve Traysâ€ in Kister Distillation Symposium 2017, Proceedings of Topical Conference, AIChE Spring Meeting, San Antonio, Texas, March 26-30, 2017.

# 513926 Operational Excellence in Refinery Distillation Better Decision Faster

**Abstract:**

**Operational Excellence** results from making better decisions, faster, and then perfectly and consistently executing those decisions. In this paper we discuss how Simulation-based unit monitoring has been successfully applied to crude distillation units. Simulation-based monitoring enhances traditional unit performance monitoring by providing governance of data quality based on fundamental physical and chemical laws. In addition the simulation provides a richer set of analytic data which is more meaningful than raw measurements (for instance exchanger fouling factors and column performance parameters, instead of raw temperatures, pressures and flows), this allows the end users to make more insightful operational and maintenance decisions. Using this type of tool successfully requires building outputs into workflows, and we therefore discuss best practices for systematically aligning Simulation results with the decisions, tasks and work processes of relevant people in the organization.

# Three Products with High Purity Dividing Wall Columns in Operation

Manish Bhargava, Roomi Kalita

GTC Technology US LLC., Houston, USA
mbhargava@gtctech.com
rkalita@gtctech.com

Dividing Wall Columns (DWC) involving less stringent product specifications have been proposed and operated in refinery operations in the past. These columns are often seen in individual columns such as naphtha splitters or reformate splitters. As the composition of product flows are flexible, these columns do not require advanced control methodology to maintain the product purity.

In recent years, DWCs are being implemented for production of high-purity aromatics such as Benzene-Toluene-Xylene (BTX) and recovery of mixed xylenes. This paper investigates the control and operation of these columns by use of a combination of plant operating data and dynamic simulation models. A grassroots middle dividing wall column is used in both of the cases under consideration. The aim is to identify the most favorable and economic method of operating such a column for the maximum profits and efficiency. The advantages of DWCs are evident. In a majority of the cases, DWCs have proven to generate 20-30% energy and capital investment savings as compared to conventional techniques. However, high purity aromatics extraction adds a new complexity to the operation of dividing wall column.

## Introduction

Aromatics production units usually consist of a two-column sequence. For a mixed xylenes recovery unit, toluene is separated in the first column followed by a xylene column. Mixed xylenes and the $C_{9+}$ cut are removed in this column. For a BTX separation, the bottoms from the extractive distillation unit are processed in a benzene column followed by a toluene column. High purity benzene, toluene and xylenes are produced in that sequence.

These high purity separations can be done in a middle dividing wall column, which has a vertical wall running through the central part of the column. Essentially two sections are created within the column. These can be symmetrical or asymmetrical based on the degree of separation required and the feed quality. The pre-fractionation afforded by the wall on the feed side helps in better separation and component recovery. The wall forces the middle boiling cut (either toluene or mixed xylenes) to concentrate on the other side of the wall with minimal adulteration from the lightest and heaviest components.

### 1.1 Mixed xylenes recovery unit installed at Tonen General, Japan

The grassroots GT-DWC® unit separates a mixed xylenes stream from the heavy reformate (Bhargava et al., 2016). Toluene and the $C_9$+ cut are the top and bottoms products. The DWC column pressure is elevated so that the overhead vapors of the column are used to heat the reboilers of the upstream columns. This presents an extra challenge to the operation of the DWC. Besides maintaining the high purity product specifications, the refiner also needs to generate sufficient vapors for the stable operation of the two upstream columns.

The TonenGeneral Group has an operational middle dividing column that recovers mixed xylenes from a full-range reformate feed (Figure 1). The new column is downstream of a pre-existing benzene extraction unit (BEU). The unit has two columns that recover light reformate, benzene and heavy reformate. In addition, the two existing columns operated on steam reboilers. The refinery, being located in Japan, had plot space constraints. Hence, the conventional two-column solution was abandoned in favor of a compact DWC.

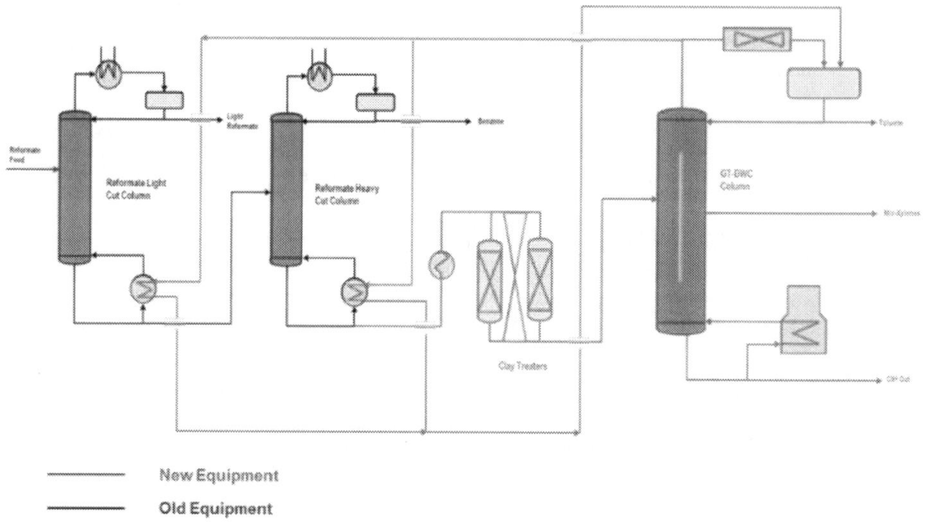


*Figure 1: New GT-DWC® mixed xylenes column was installed downstream of the existing light and heavy cut columns in the Benzene Extraction Unit (BEU)*

In spite of the complexities, the operation and control of a dividing wall column is fairly simple. The control philosophy for the column is shown in Figure 2.

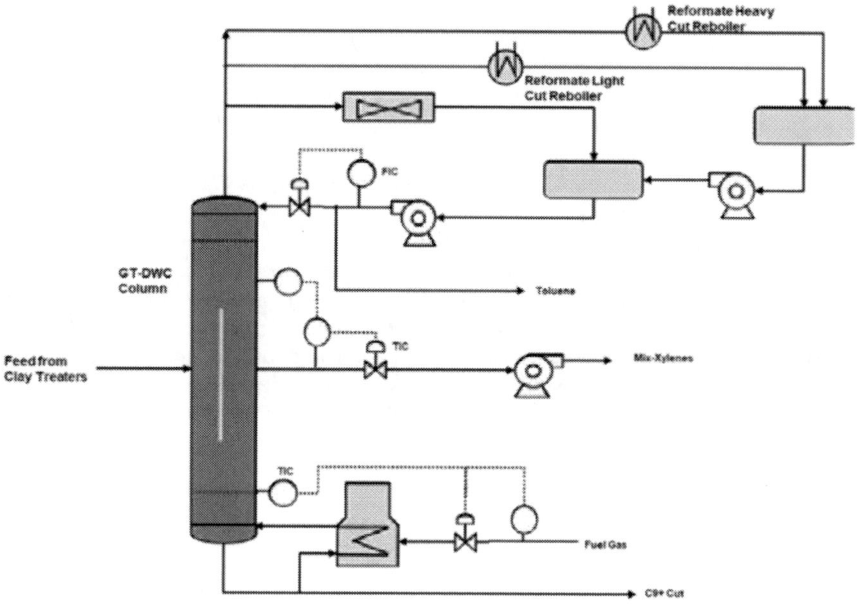


*Figure 2: Control Scheme of the GT-DWC® mixed xylenes column*

Since the DWC is a three-cut column, the column is defined by three primary control loops. These include the following:

1. The column operates on a fixed reflux rate. Fixing the reflux rate helps to generate consistent vapors for the upstream reboilers and thereby, the product quality of the two columns remains stable.

2. Since the column requires a fixed reflux rate, the heat duty input of the DWC column needs to be defined. This is done by cascading the fuel gas flow of the fired heater to the column bottoms temperature.

3. The column produces high purity mixed xylenes. This is achieved by cascading the side-cut product rate to one of the upper tray temperatures.

*Figure 3a: High Purity Mixed Xylene Column (Operational since March 2016);*
*Figure 3b: Off-center Dividing Wall*

The DWC is equipped with a combination of proprietary internal and external liquid splitting devices. The column maintains operational flexibility for drastic change in feed composition or product specs by utilizing this extra degree of control in such scenarios.

Figure 4 below provides the test run data of mixed xylenes product. $C_8$ aromatics product purity containing ethylbenzene is maintained at around 99 wt%. The recovery rate of $C_8$ aromatics at plant start-up is around 80 to 90 wt%. After operational adjustments and controller optimization, TonenGeneral currently maintains a high recovery rate of 95-99 wt%.

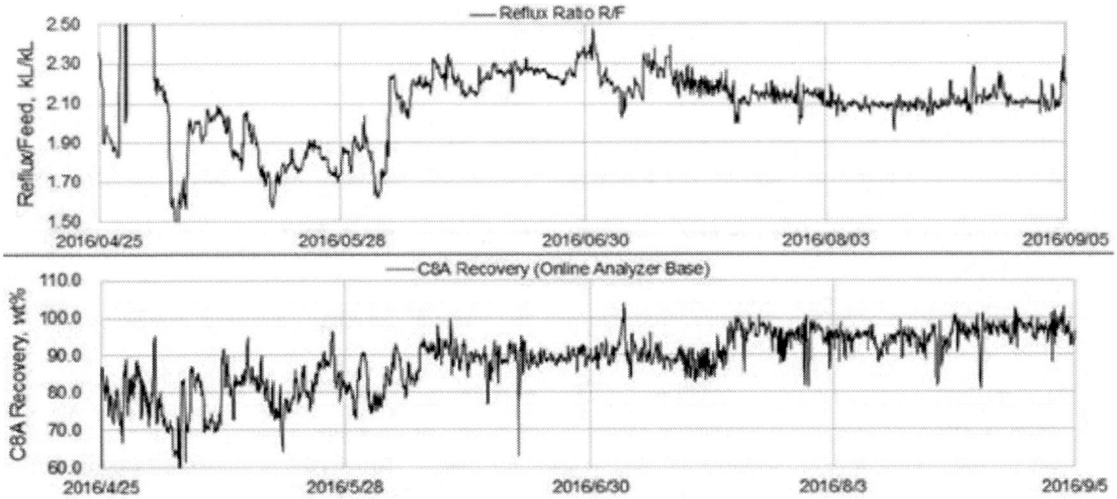


*Figure 4: Plant operating data of the GT-DWC® mixed xylenes column*

Since its start-up, the DWC unit maintained a consistent product specification as shown in Table 1.

*Table 1: Product Specification of Mixed xylenes*

| Components | Units | Specifications | Remarks |
|---|---|---|---|

| | | | |
|---|---|---|---|
| $C_8$ aromatics | wt% | 98.5 | Minimum |
| Ethyl Benzene | wt% | 18.0 | Maximum |
| Non-aromatics | wt% | 1.0 | Maximum |
| $C_9$ aromatics | wt% | 0.3 | Maximum |

**1.2**

### 1.3 Benzene-Toluene-Xylene separation unit

A grassroots middle dividing column producing high purity benzene, toluene and mixed xylenes is under construction in China for Dongying Yatong Petrochemical Co. Ltd. The startup of this column is planned for March 2018.

The column will recover benzene and toluene products in a single column from an aromatics rich extract. For operational flexibility, the column uses a combination of external and external liquid splitting. The three products required in this particular column has precise specifications and hence, the column controls need to be very specific to regulate the products.

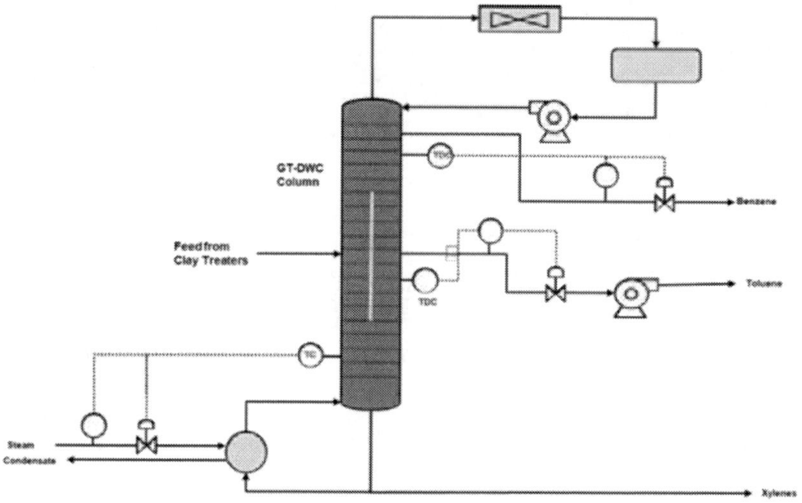


*Figure 5: Control Scheme of the GT-DWC® BTX column*

The column will utilize a simplified control scheme to produce on spec products as shown in Figure 5.

First, the column produces high purity benzene as the top product. To maintain the strict product specifications, the benzene product rate is cascaded to the differential temperature on the upper trays of the column.

Next, the toluene product flow is cascaded to the differential temperature of trays below the side cut on the product side of dividing wall.

Last, the steam flow rate to thermosyphon reboiler is cascaded to the column bottoms temperature.

To validate the control scheme, a dynamic simulation model has been developed in Hysys. The simulation model predicts the performance of the column based on several factors including column top pressure, reflux drum level, column bottom temperature, column sump level and the bottoms product flow. The feed flow and composition were varied to observe the shift in column behavior. The simulation model has been adjusted to match the steady state simulation results (Table 3) as closely as possible.

The model simulations are based on a four-column system to mimic the operation of the different sections in a middle DWC. Additionally, conservative estimations have been made to account for the pump characteristics and control valve pressure drops.

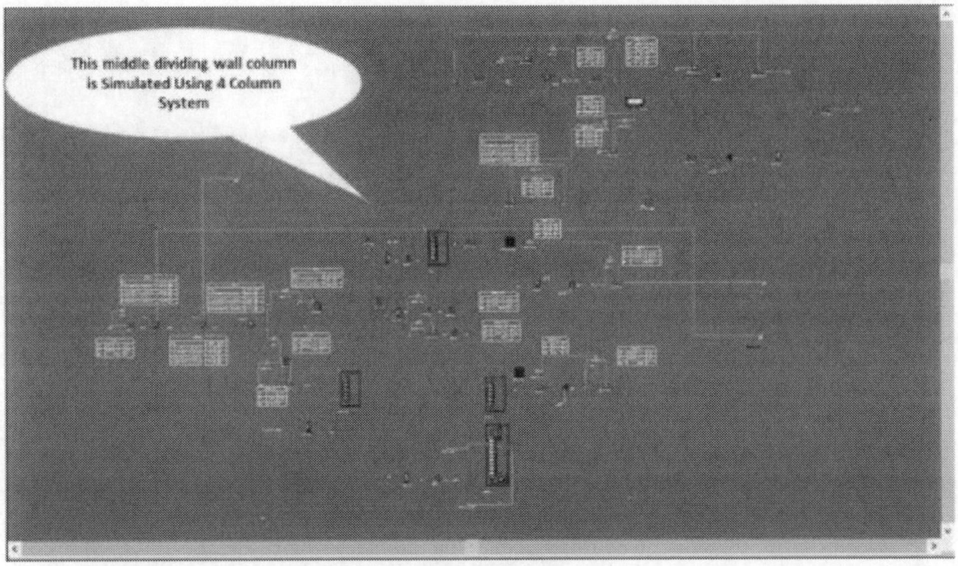


*Figure 6: Four-Column Dynamic Simulation Model of BTX DWC Column in Hysys*

The control scheme was validated to achieve the product specifications outlined in Table 2.

*Table 2: Product Specification of GT-DWC BTX Column*

| Components | Units | Specifications | Remarks |
|---|---|---|---|
| Benzene Product | wt% | 99.95 | Minimum |
| Toluene Product | wt% | 99.920 | Minimum |
| Xylene Product (IBP/FBP) | °C /°C | 137.5 – 141.5 | |

Dynamic simulation model studied the behaviour of column controls by varying the feed rate from -10% to10%. The results are presented in Figure 6a and 6b.

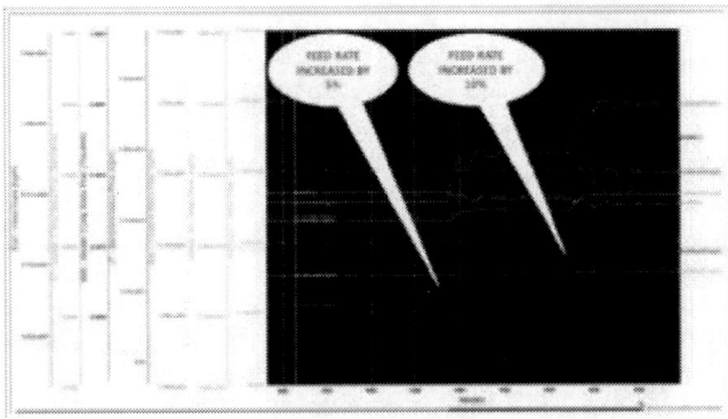

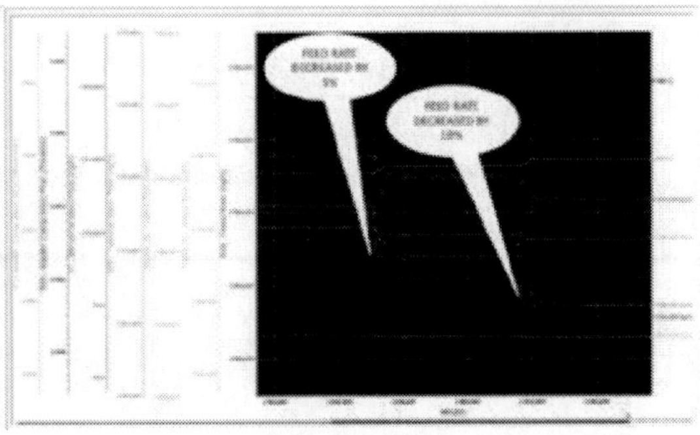

*Figure 7a: Dynamic simulation response for step change in feed increase of 5% and 10%;*
*Figure 7b: Dynamic simulation response for step change in feed decrease of 5% and 10%*

The dynamic simulation also validated the control scheme behaviour for step change in feed compositions. The results are presented in Figure 8.

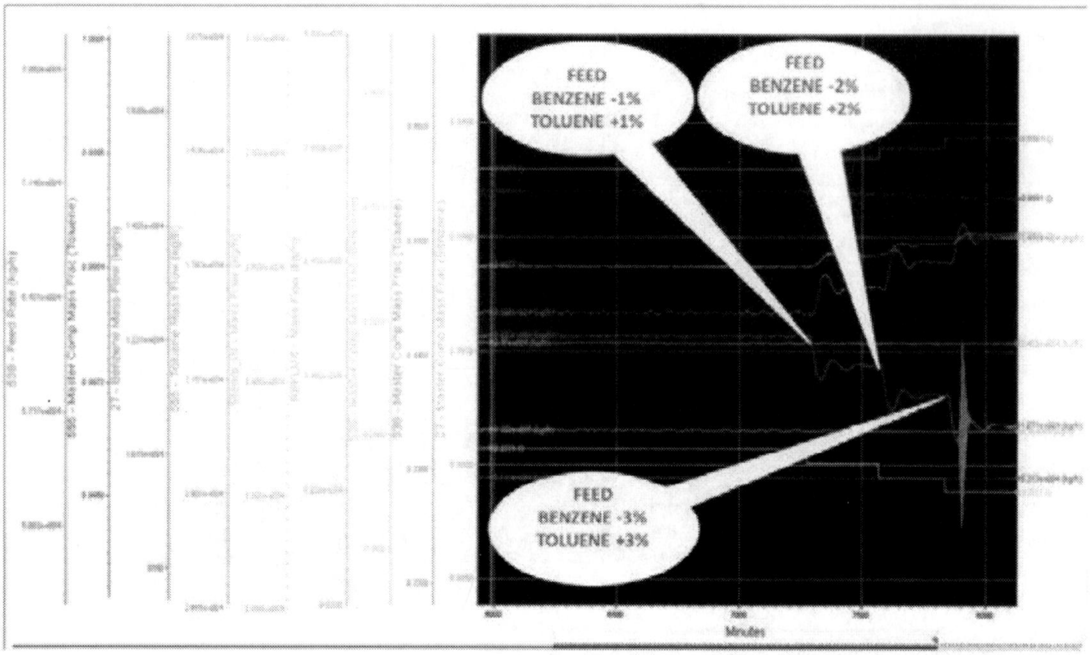


*Figure 8: Dynamic simulation response for step change in benzene and toluene composition in Feed.*

The dynamic model corroborates the design control scheme for the DWC column. The column generates high purity benzene and toluene products even with a +/- 10% variation in the design flow. The product purities are maintained even when the benzene/toluene composition in the feed varies by about +/-3%. The dynamic response for all the control loops evaluated is normal and as per expectations of the design.

This paper will further provide the comparison of operating data with simulation results after the start-up due in March 2018.

## Conclusions

Much research has been done on Dividing Wall Columns till date in conjunction with successful implementation in several refinery applications. Despite this, a lot of refiners are apprehensive of using these innovative columns in high-purity separations. Contrary to prevalent belief, DWCs are similar in operation to regular distillation columns. The product quality in these columns can be maintained by using a simple three-control methodology. Plant operating data and simulation model results endorse the use of these controls.

### References

Bhargava M., Kalita R., Gentry J.C., Suzuki N., 2016, Improved Distillation Efficiency, PTQ (October)

Bhargava M., Kalita R, Gentry J.C., 2017, Dividing Wall Column Provides Sizable Energy Savings, Chemical Processing (May)

## 513035 Operable Version of Dividing-Wall Columns: A Control Perspective

**Abstract:**

Distillation is one of the most energy-intensive unit in chemical process industries. Conventionally, ternary mixtures are separated in two distillation columns by separating the most volatile component first (direct-split sequence), separating the least volatile component first (indirect-split sequence), or distributing the components (mid-split sequence). Columns in these sequences can have separate reboilers and condensers, or they can be thermally coupled to share liquid and vapor and operate with a single reboiler and condenser. The latter type of distillation columns are called Petlyuk columns. The most widely implemented dividing-wall column (DWC) provides capital and energy savings up to 40% compared to the conventional columns. DWC, is a single shell column with top, middle and bottom sections, where the middle section is divided into prefractionator and main sections by inserting a vertical wall. This arrangement allows the operation of the Petlyuk column in the DWC.

Despite the thermodynamical and economical advantages, a major drawback of DWCs that prevent them from being widely implemented in industries is the controllability issue that arise due to vapor split at the bottom of the DWC. The vapor split is proportional to the cross-sectional area of the prefractionator and the main columns, and cannot be manipulated during the operation because of the fixed physical location of the wall. Not only this results in the loss of a degree of freedom for controlling the operation of the DWCs, disturbances in the vapor split due to the external disturbances may not be well understood during the simulations or rejected well by the control systems during the operation.

Ramapriya et al.[1] proposed a method to synthesize thermodynamically equivalent DWCs by extending the middle-wall to the top and/or bottom of the DWC, adding additional condenser and/or reboiler, and transferring mass between the two sides of the wall through liquid flow transfer. This method can replace the vapor split at the bottom of the DWC with a liquid only transfer which is available for manipulation for control purposes. In this work, we present the control advantages of DWCs with extended wall to the top and/or bottom by considering the ternary separation of benzene, toluene and ortho-xylene in a DWC. We demonstrate the results by developing a new PID control strategy and study the responses of the column to disturbances in flow and composition of the feed. The DWC with extended wall not only can be practically operated, it can appropriately reject the external disturbances and deliver on-spec products.

References:

1. Madenoor Ramapriya, G., Tawarmalani, M. and Agrawal, R. (2017), A systematic method to synthesize all dividing wall columns for *n*-component separationâ€"Part I. AIChE J. doi:10.1002/aic.15964

## 512312 Using a Dividing Wall Distillation Column to Isolate Trace Impurities

**Abstract:**

As part of an ongoing experimental study of dividing wall columns at The University of Texas at Austin, the control of a fixed design DWC isolating a middle-boiling impurity will be presented. The pilot plant column is six inches in diameter and approximately 30 feet tall. The column has 24 theoretical stages of structured packing and a welded wall in the horizontal and vertical middle of the column. A fourth trace component was added to the feed to mimic industrial systems in which trace impurities are present. Disturbance testing of the pilot column included changes in feed composition and operating variables such as the overhead reflux and the wall split. Experimental results were used to validate a comprehensive rigorous dynamic model that includes heat transfer and pressure drop calculations.

# Network of Dividing Wall Columns in Complex Process Units


Manish Bhargava, Roomi Kalita

GTC Technology US LLC., Houston, USA
mbhargava@gtctech.com
rkalita@gtctech.com



The majority of dividing wall column (DWC) applications in the refining industry consists of stand-alone columns, whether a retrofit or a grassroots column. Functional DWCs are commonly encountered in naphtha splitters or reformate splitters. However, the technology can potentially improve the efficiency and profitability of complex refinery units as well. One such area is complex processes like naphtha hydrotreating (NHT) and isomerization (ISOM), which generally consists of a network of columns. This article discusses how DWC technology is incorporated to combine two or more columns in these processes to improve overall profitability.

In addition, these units often operate at high pressures and temperatures that result in a costly and energy-intensive operation. With the demand for such units ramping up in the refining industry, DWCs can revolutionize these process schemes. DWCs not only improves energy efficiency of the processes, but can also provide a host of other benefits. These include lower capital investment and relatively smaller plot area as compared to a full-fledged conventional scheme.


## Introduction

Typically, in the NHT Unit (Figure 1), after the reactor section, the stabilizer column removes the non-condensable gases from the feed. The top liquid product goes to a deethanizer column to recover LPG (liquefied petroleum gas) product. The stabilizer bottoms are fed to a two-column naphtha splitter sequence separating a light naphtha (mostly $C_5$-$C_6$ components), mid-cut naphtha ($C_7$) and heavy naphtha ($C_8$ and heavier).The light naphtha is processed by the deisopentanizer to separate the i-$C_5$ rich stream at the top. The n-$C_5$ and heavier fraction is then fed to the ISOM reactor.

In most process schemes, the stabilizer operates at a high pressure, which necessitates the use of relatively expensive medium pressure (MP) steam. Additionally, due to the use of partial condensation, significant $C_3$-$C_4$ losses are observed in the off-gas. This leads to the use of a supplementary deethanizer column to recover LPG from the off-gas. The isomerization feed is prepared from stabilizer column bottoms in a series of one or more naphtha splitting columns.

Similarly, an ISOM unit (Figure 1) consists of a multitude of columns including deisopentanizer, stabilizer, dependanizer and deisohexanizer columns. The deisopentanizer column separates high octane IC5 components from feed and recycle streams. The low octane components from the deisopentanizer bottoms are sent to a isomerization reactor to produce high octane components along with some lights. The stabilizer removes the lighter hydrocarbons ($C_{4-}$) in the off-gas. The stabilized isomerate is fed to a depentanizer column to concentrate $C_5$s. The downstream deisohexanizer column then separates a light isomerate (mainly i-$C_6$) and a heavy isomerate (mainly $C_{7+}$ cut) along with a concentrated n-$C_6$ cut. The n-$C_6$ cut is recycled to the ISOM reactor for octane upgrading.

These series of columns present a good opportunity to incorporate DWC technology to improve the efficiency of the whole process. DWCs can be utilized to combine the operation of two or more columns while achieving the same product specifications as the conventional scheme. As a result, capital investment is lowered by 20-30%. Moreover, these columns can either be operated at lower temperature utilities or have lower utility requirements. Energy savings in the range of 20-30% is expected with a DWC configuration.

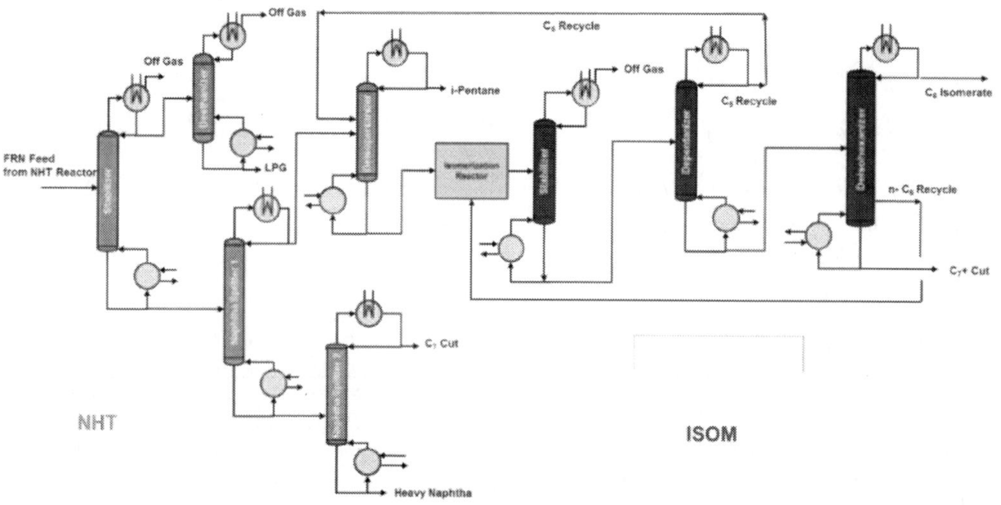


*Figure 1: Conventional Naphtha Hydrotreating and Isomerization Process Scheme*

**Dividing Wall Columns in Naphtha Hydrotreating (NHT) Units**

In the NHT unit, as shown in Figure 2 and Figure 4, the stabilizer can be replaced with a top dividing wall GT-LPG MAX® column. The wall completely isolates the top section of the column into halves that basically operate as independent columns. The wall prevents any intermixing or leakage between the two sides. As a result, separate unit operations can be performed within the column. In this case, one section is used for absorption and the other for distillation. In addition, the parallel separation zones that have been created tend to minimize the overall height of the column. A similar number of theoretical stages (or trays) can be accommodated in the two zones for the desired separation.

A heavy hydrocarbon stream strips the off-gas of $C_3$-$C_4$ components on the absorption side of the column. These components are concentrated and removed on the distillation side as LPG product. A portion of the bottom product from the GT-LPG MAX® column can be used for absorption if the feed contains a suitable amount of $C_5$s. Alternatively, a lean naphtha stream can be used either alongside the heavier $C_5$ stream or independently as the absorption medium.

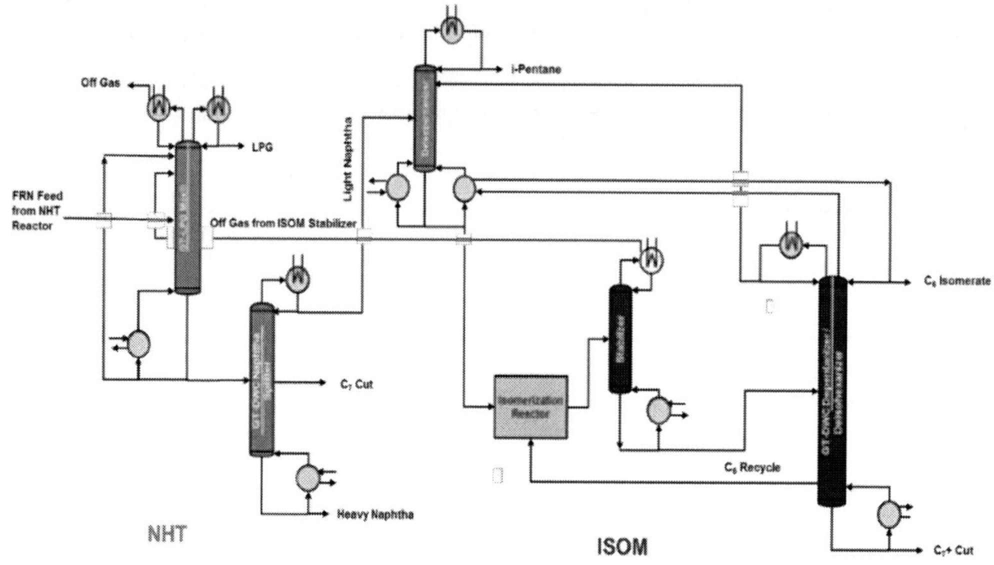


*Figure 2: Heat integrated GT-DWC® Process Scheme for NHT and ISOM units*

Other off-gas streams in the refinery that are rich in $C_3$-$C_4$ components, such as the ISOM stream, can be fed to this DWC column to improve LPG recovery (Figure 2 and 3). Consequently, other stabilizers can be operated under relatively lower pressures, and lower temperature heating utility.

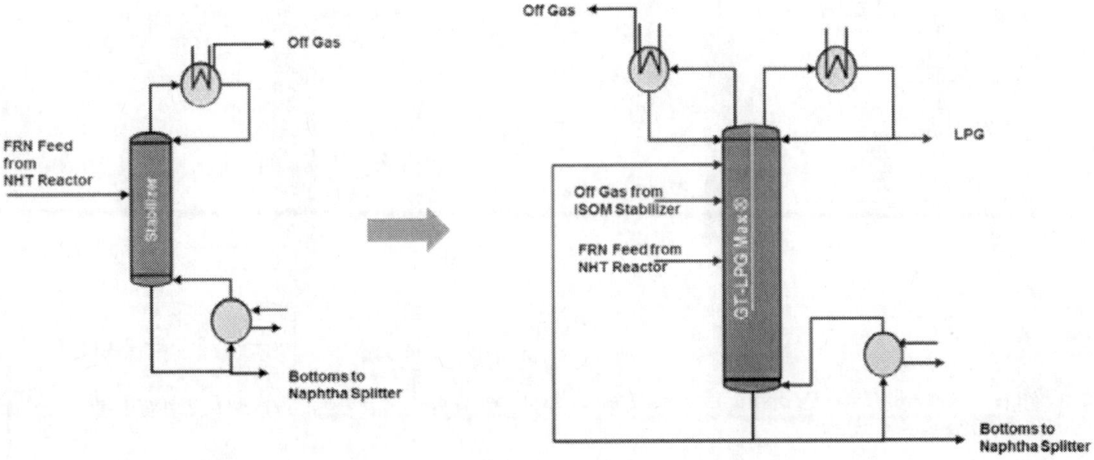


*Figure 3: Conventional Stabilizer v/s GT-LPG MAX® Column*

The two naphtha splitters in the conventional sequence are replaced by a GT-DWC® naphtha splitter. This column has a middle dividing wall column (Figure 4). DWCs work on the principle of removing intrinsic thermodynamic design flaws in conventional distillation columns. One of these flaws arises due to the back-mixing of the feed with the side-cut based on the location of the two streams. The side-cut quality is affected by the contamination of lighter or heavier components. Using a dividing wall column eliminates the problem and produces a better quality side-cut. Hence, the two naphtha splitter operation is possible in a single DWC.

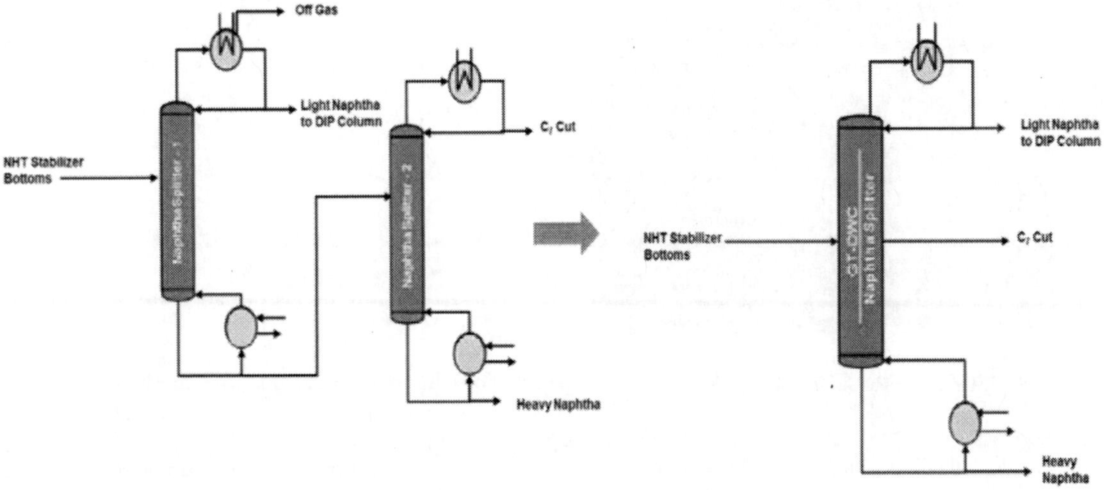


*Figure 4: Conventional Naphtha Splitters v/s GT-DWC® Naphtha Splitter*

**Dividing Wall Columns in Isomerization (ISOM) Units**

A top dividing wall concept can also be applied to combine the operation of the depentanizer and deisohexanizer columns in the ISOM unit (Figure 6). One side of this GT-DWC® depentanizer/deisohexanizer column acts as a pre-fractionation column to concentrate the $C_5$s components. The middle boiling components ($C_6$s) and the heaviest boiling components ($C_7$s and heavier) are pushed down the column to the main fractionation side. Here, $C_6$ isomerate is separated as the top product. A concentrated n-$C_6$ stream is removed as side-cut from below the dividing wall section and recycled to the ISOM reactor. $C_7$+ cut is obtained as the bottom product of this column.

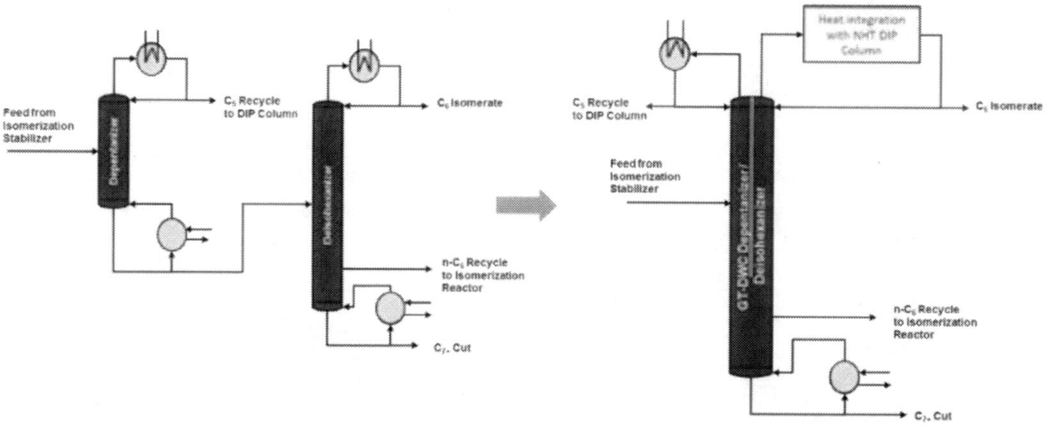


*Figure 5: Conventional Depentanizer and Deisohexanizer Columns v/s GT-DWC® Depentanizer / Deisohexanizer Column*

*Figure 6: A Top Dividing Wall Column in an ISOM unit in operation at BPCL refinery (Mumbai, India).*

For regions having high utility costs, further energy savings are possible through the overhead heat integration of a top dividing wall column. The GT-DWC® column operates at an elevated pressure. As a result, the hot overhead vapors on the main-fractionation side can be used to provide heating duty to the upstream deisopentanizer column in the NHT unit (Figure 5 and 6). The stream is capable of providing the entire duty required by the deisopentanizer column. The column otherwise operates on LP steam typically.

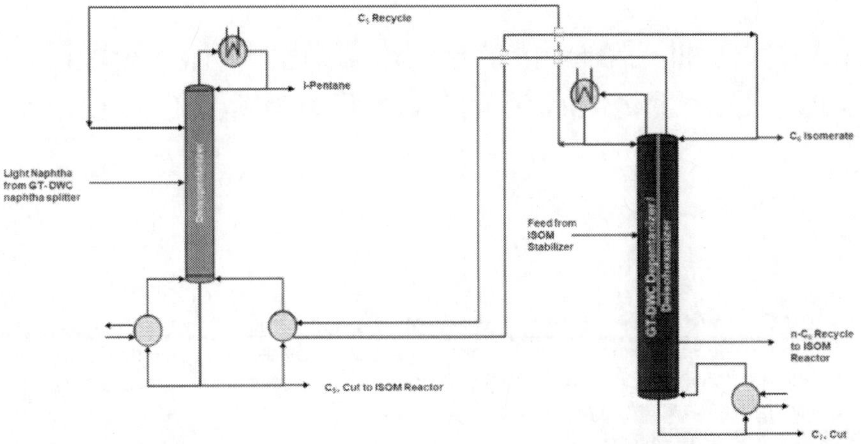


*Figure 7: Overhead Heat Integration between Deisopentanizer and GT-DWC® Deisohexanizer Column*

The benefits of a network of Dividing Wall Columns in NHP and ISOM units is summarized in table 1 & 2.

*Table 1: Equipment Count, CAPEX and OPEX Comparison*

| Parameters | Units | Conventional Design | GT-DWC® Design |
|---|---|---|---|
| No. of Columns | - | 8 | 5 |
| Energy Savings | % | - | 24 % |
| Capital Cost | % | Base | 70% of Base |

*Table 2: Heating Utility Comparison*

| Parameters | Units | Conventional Design | GT-DWC® Design with Heat Integration |
|---|---|---|---|
| Feed Rate to NHT Unit | TPH | 57 | 57 |
| Isomerate product | TPH | 28.6 | 28.6 |
| RON of isomerate product | - | 92 | 92 |
| LPG Product Rate | TPH | 5.0 | 7.5 |
| Off Gas Rate | TPH | 3.7 | 1.2 |
| NHT Stabilizer | MMKcal/hr | 7.0 | 7.3 |
| Deethanizer | MMKcal/hr | 2.0 | - |
| Naphtha Splitter 1 & 2 | MMKcal/hr | 16.6 | 13.3 |
| Deisopentanizer | MMKcal/hr | 12.7 | 0.9 |
| ISOM Stabilizer | MMKcal/hr | 3.0 | 3.0 |
| Depentanizer | MMKcal/hr | 8.0 | 20.9 |
| Deisohexanizer | MMKcal/hr | 10.4 | - |
| Total Heating Duty | MMKcal/hr | 59.7 | 45.4 |

**Conclusion**

Dividing Wall Columns, merging the operation of two or more sequential columns into a single DWC, provide an innovative method of reducing the capital and energy costs of complex refinery processes such as isomerization and naphtha hydrotreating.

Integrating DWC technology into a NHT/ISOM process scheme can provide substantial benefits to the client:

- Lesser number of columns and associated equipment for the whole configuration
- Improved LPG recovery
- Reduced energy costs due to low temperature utility used for heating
- Better heat integration within the columns.

# Use of Dividing Wall Columns to Increase Throughput and Improve product Specifications in Two-Diameter Columns

Manish Bhargava, Roomi Kalita

GTC Technology US LLC., Houston, USA
mbhargava@gtctech.com
rkalita@gtctech.com

Dividing Wall Columns (DWC) provide a new approach to age-old distillation techniques. These columns have become popular lately owing to the energy and capital benefits that they provide. These columns are also quite versatile and flexible in their applications. Currently, DWCs are found in both retrofit and grassroots applications for reformate splitters, naphtha splitters amongst others.

DWCs work on the principle of removing thermodynamic design flaws in traditional distillation columns. These limitations are encountered in both two-column separation sequences as well as side-cut columns, especially columns with two-diameter sections. This paper highlights these inherent design flaws and illustrates how DWC technology helps to workaround these flaws to generate a more efficient and robust column. Better product qualities and reduced energy consumption are a direct result of such modifications. Occasionally, DWC technology can provide added benefits wherein a column is freed up from a separation sequence.

Two examples of revamp of naphtha splitter columns are presented wherein middle DWC technology has been implemented. In one case, a DWC column produces better product specifications than the original design. In the other case, a two-column sequence is replaced by a single DWC column for the same product quality. This results in freeing up the second column for use in another service. The two columns retrofitted as a DWC have two-diameter sections and are operational in one of the largest refineries in South East Asia.

## Introduction

Conventional distillation of three products is generally carried out in a sequence of two columns. In some separations, a side-cut column is used. Both two-column separation and side-cut columns have certain thermodynamic inadequacies. These are explained below:

First, let us consider a two-column separation sequence (Figure 1a). Here, the first column removes the lightest key (A). The middle cut (B) and the heaviest key are removed at the top and the bottom of the second column in that order. However, the middle boiling components reach a concentration peak somewhere near the bottom of the first column. At the same time, the concentration of C increases in this area leading to a remixing of the two components (Figure 1b). The same components are then separated in the next column utilizing additional heating duty. The separation achieved in the first column using heating duty is somewhat wasted. As a result, conventional two-column separation sequences generally see high reboiling and cooling duties.

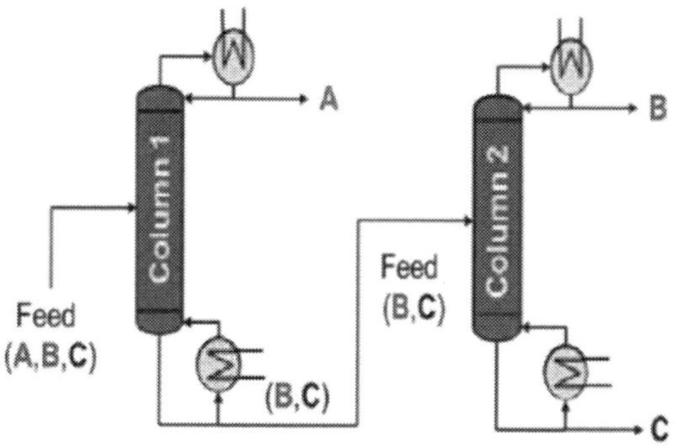


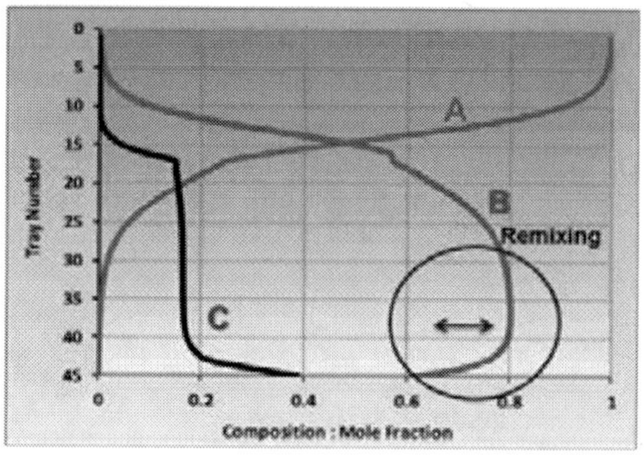

*Figure 1a: Conventional Two-Column Separation; 1b: Composition profile of components in a Two-Column Separation.*

Next, a side-cut column (Figure 2a and 2b) also faces a similar thermodynamic problem. Here, the feed often intermixes with the side-cut affecting the product quality. For side-cuts below the feed nozzle, a lot of heavier components end up in the middle cut affecting the final boiling point (FBP) of the product. Likewise, a side-cut above the feed nozzle has problems meeting the initial boiling point (IBP).

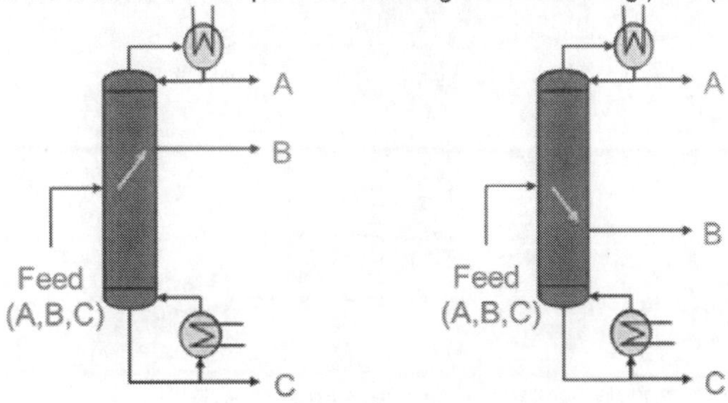


*Figure 2a: Thermodynamic Inadequacies in Side-Cut Column for a Side-Cut above the feed; 2b: Thermodynamic Inadequacies in Side-Cut Column for a Side-Cut below the feed.*

A dividing wall column (Figure 3) successfully eliminates these inadequacies and makes the whole separation more efficient while minimizing the utility requirements for the same separation. This is done with the aid of a middle dividing wall design. The wall is located somewhere in the middle of the column. As a result, the column is split up to a pre-fractionation side and a main fractionation side. The wall removes the possibility of the feed mixing with the side-cut leading to a better product quality. Due to the pre-fractionation area available, the middle boiling components are distributed amongst the lightest and the heaviest key based on the degree of volatility. These components are eventually concentrated and removed in the center of the main fractionation side. A DWC is hence able to generate a better product quality in a single column. The heating requirements are also lowered by about 20-30% due to the added advantage of pre-fractionation.

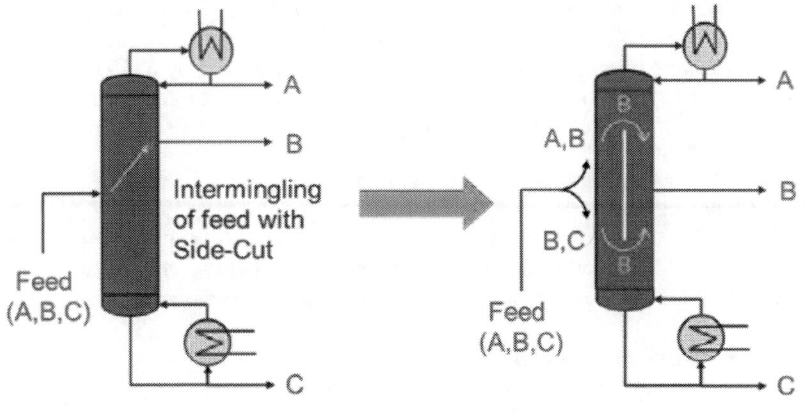


*Figure 3: Conventional Side-Cut Column v/s GT-DWC Column*

**DWC for better product specification in a single column:**

DWC technology has been used to retrofit a FCC unit naphtha splitter column (Figure 4a). The unit in question separates cracked naphtha into three products – namely light naphtha, heart-cut naphtha and heavy naphtha. The light and heavy products are routed for gasoline blending, while the heart-cut naphtha feeds a reforming unit. For higher para-xylene production in the reforming unit, the heart-cut naphtha should contain a good amount of aromatic-rich streams (primarily $C_8A$ and $C_9A$ precursors).

The column has two reboilers. The side reboiler operates on a low temperature utility, LCO (Light Cycle Oil). The bottom reboiler operates on a high temperature utility, HCO (Heavy Cycle Oil).

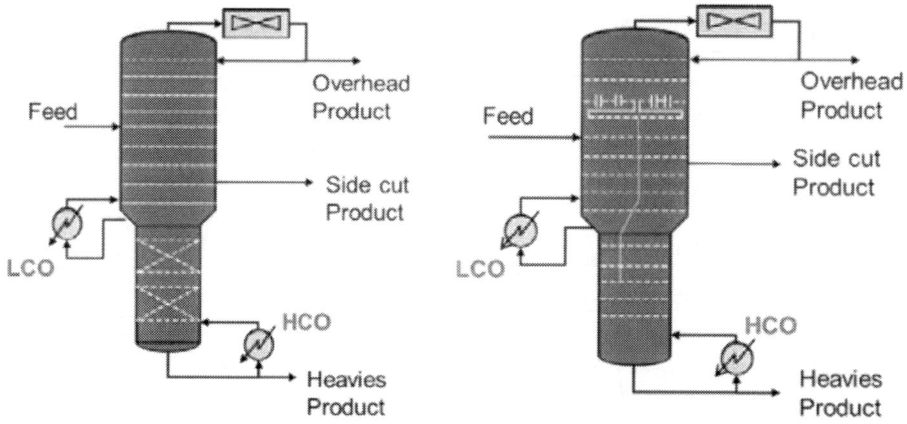


*Figure 4a: Original Design of FCC Unit Naphtha Splitter; 4b: After revamp to GT-DWC®*

In the original design, the side-cut location is located below the feed tray location. This maximizes the reboiling duty of the side reboiler, while minimizing the bottom reboiler duty. The lower reboiler duty cannot be increased much due to the smaller diameter section. However, this eventually led to a contamination by heavier components in the heart-cut affecting its quality. The heart-cut naphtha FBP ended up in the range of 190°C-198°C. This adversely affects the downstream reformer unit. Thus, one of the main objectives of the revamp is to lower the FBP of the heart-cut naphtha to 180°C to meet the reformer feed specifications.

One of the common approach to such a revamp is to install an additional column to process the heart-cut naphtha. However, as mentioned earlier, a two-column separation sequence requires higher heating and reboiling duties. On top of that, the new column requires additional capital investment for the extra equipment needed. So, the refiner narrowed down upon DWC technology to revamp the column.
The naphtha splitter column is retrofitted with a middle dividing wall column (Figure 4b). The wall is placed slightly off-center such that the side reboiler is confined to the pre-fractionation section. The lightest components are moved towards the top of the column using maximum duty from the LCO reboiler. A portion of the wall also extends to the smaller diameter section. This prevents the lightest components from moving downwards without greatly affecting the duty of the bottom reboiler. The heart-cut naphtha and the heavy naphtha is separated using the HCO reboiler. The wall eliminates the back-mixing of the feed with the side-cut while optimizing the duties on the reboiler. As a result, a better quality heart-cut naphtha is separated in the same column.

For this revamp, the capital investment is minimal as most of the modifications are confined to the column. The associated equipment are not touched. Table 1 shows the heart-cut naphtha product quality before and after the revamp. Following test runs after the start-up of the column, the overlap between the heart-cut and the bottom cut reduced from 37°C to -3°C. The overlap is defined as the difference between heart-cut naphtha D86 95% and heavy naphtha D86 5%. A smaller overlap indicates a good separation between the two products.

*Table 1: Heart Cut Naphtha Specifications before and after revamp*

| Parameters | Units | Naphtha Splitter before Revamp | Naphtha Splitter after Revamp to DWC |
|---|---|---|---|
| Heart-cut Flowrate | tonnes/hr | 180.4 | 151.0 |
| D86 (IBP/FBP) | °C | 110.5-190.6 | 110.5-172 |
| Overlap (Heart-Cut Naphtha D86 95% - Heavy Naphtha D86 5%) | °C | 37 | -3 |

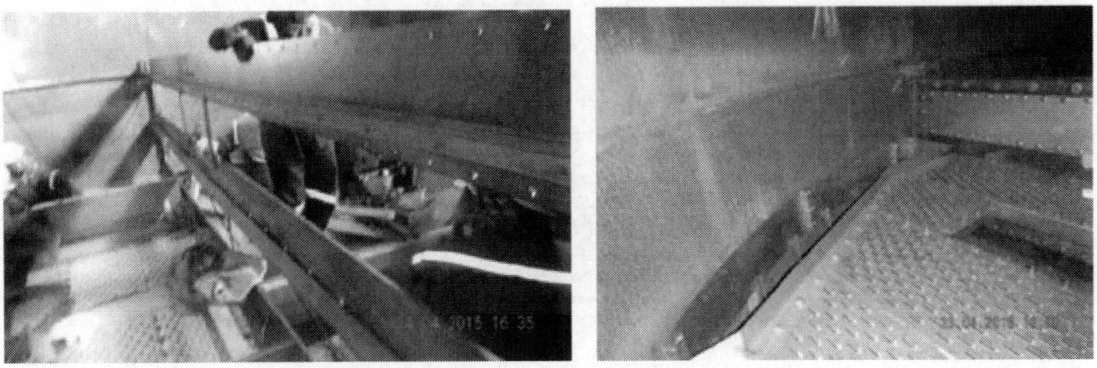

*Figure 5a: Installation of the dividing wall in FCC Naphtha Splitter; 5b: Dividing wall internals inside the tower*

**Two-Column Application to a Single Column Application:**

In another revamp, DWC technology was employed to produce same product specifications in a single column as opposed to a conventional configuration of two-columns. The original design (Figure 6) includes two naphtha splitters columns, naphtha splitter-1 is followed by naphtha splitter-2. Column 1 separates the feed into light naphtha and heavy naphtha at the top and bottom respectively. Additionally, a side-cut is removed which acts as the feed to the naphtha splitter-2. This column separates a heart cut naphtha stream at the top of the column. The remaining heavy naphtha is removed at the bottom.

The naphtha splitter-1 column is also a two-diameter column with two reboilers. The side reboiler operates on LCO while the bottom reboiler operates on HCO. Similar to the previous example, the side-cut is removed from below the feed tray location. The naphtha splitter-2 column is a single diameter column, with two bottom reboilers. A combination of HCO and high pressure (HP) steam provides the heating duty to this column. Additionally, the column has a low pressure (LP) steam generator in the overhead system. On the whole, the two columns combined requires a substantial amount of energy to produce high-quality heart cut naphtha.

The refinery is located in Asia, where utility costs are particularly high. One of the main objectives of the revamp was to lower the utility consumption of the two columns to improve profits. Furthermore, the refiner had plans to use the naphtha splitter-2 in a different service. Thus, another objective is to produce heart cut naphtha in the first column.

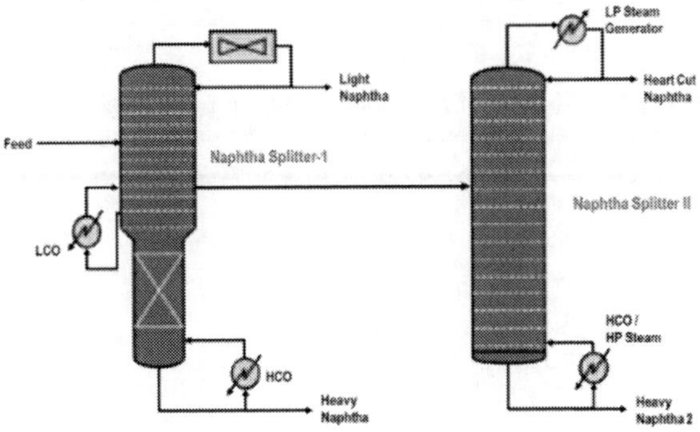


*Figure 6: Original Design of Naphtha Splitter-1 and Naphtha Splitter-2*

For the revamp, naphtha splitter-1 column is converted into a middle dividing wall column (Figure 7). The column produces a better quality heart-cut by following the same principles as the FCC unit naphtha splitter column. Thus, naphtha splitter-2 is removed from the service.

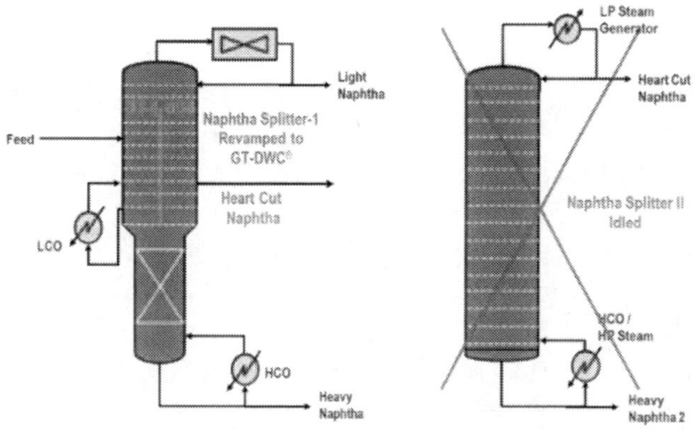


*Figure 7: Naphtha Splitter-1 and Naphtha Splitter-2 after revamp using GT-DWC®*

*Table 2: Heart Cut Naphtha Specifications before and after revamp*

| Parameters | Units | Naphtha Splitter before Revamp | Naphtha Splitter after Revamp to DWC |
|---|---|---|---|
| Heart-cut Flowrate | tonnes/hr | 165.0 | 165.0 |
| D86 (IBP/FBP) | °C | 110.6-170.0 | 110.0-170.0 |

Besides being idled from service, naphtha splitter-2 also has a HP steam reboiler in the original design. The new GT-DWC® column does not require HP steam for heating. Hence, high cost HP steam is completely eliminated after the revamp. In general, utility requirements are lowered by about 25% (Table 3), despite the new design not providing LP steam generation at the top. The product recoveries are slightly better than the original design and all equipment modifications are confined to the column.

*Table 2: Performance of Naphtha Splitter-1 and Naphtha Splitter-2 before and after revamp*

| Parameters | Units | Conventional Design | GT-DWC® Design |
|---|---|---|---|
| Feed Rate | tonnes/hr | 308.0 | 308.0 |
| Heart-cut Naphtha Product Rate | tonnes/hr | 165.0 | 165.0 |
| C8/C9 recovery in Heart-cut Naphtha | wt% | 95.0 | 98.0 |
| Total Heating Duty | MMKcal/hr | 38.0 | 28.0 |
| Naphtha Splitter-1 | | LCO (Side Reboiler) | LCO (Side Reboiler) |
| | | HCO (Bottom Reboiler) | HCO (Bottom Reboiler) |
| Naphtha Splitter-2 | | HCO/HP Steam | N/A |

In both the cases, flexibility in operation is maintained by the use of a proprietary liquid splitting arrangement (Figure 8a). In the event of a drastic change in feed, the device enables smooth and easy operation. For all DWCs, a combination of external and internal liquid splitting devices exist on either side of the wall. The mechanism is facilitates via gravity and hence, does not require an additional pump or drum. Moreover, the resultant external split piping diameter is low and requires the use of low pressure drop control valves.

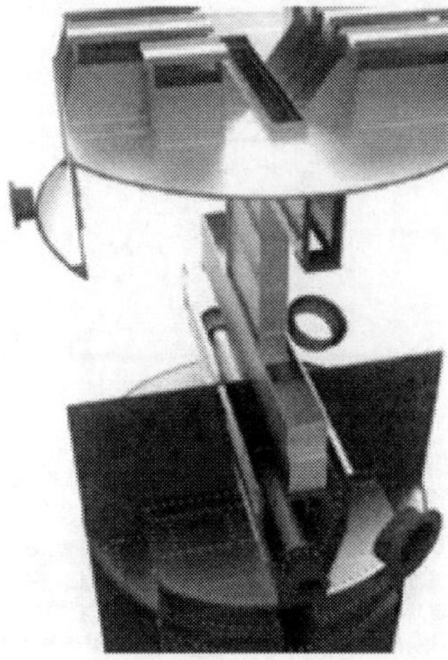

*Figure 8a: Liquid Splitting Distributor in DWC column; 8b: Top View of a four-pass Dividing Wall Column Tray Internal*

## Conclusions

Dividing Wall Columns has been successfully implemented in two units with unique outcomes. In both cases, DWCs has benefitted the refinery in a variety of ways, including

- Better product quality of the middle cut in a single column with the same utility requirements.
- In another scenario, similar product specifications are achieved with a 25% reduction.
- A whole column is idled in the second revamp by making the same product in one column instead of the original two columns.
- For both cases, the equipment modifications are limited to the column internals, thereby minimizing the need for new associated equipment.
- Last, proprietary liquid split arrangement provides an economical and flexible mode of control for future deviations in feed conditions.

## References


- Bhargava M., Kalita R, Gentry J.C., 2017, Dividing Wall Column Applications in FCC Naphtha Splitter Columns, Hydrocarbon Processing (September)
- Bhargava M., Kalita R, Gentry J.C., 2017, Converting Two Columns into One, PTQ (October)

# 512690 A Rate-Based Equation-Oriented Parallel Column Model: Application to Dividing Wall Columns

**Abstract:**

The literature on Dividing Wall Columns (DWCs) has been growing extremely rapidly of late. Indeed, it is already sufficiently large that several review articles have been published [1-3].

With very few exceptions, simulations of DWCs are carried out using an interlinked multi-column *equilibrium stage model* in a commercial simulator. (A notable exception is the work of Mueller and Kenig [4-5] who used a rate-based model to describe reactive distillation in dividing wall columns. Also noteworthy (and unusual) about their work is that they used Aspen Custom Modeler, a commercial tool that employs equation-based solution techniques. This is in complete contrast to all other models (when they have been described in sufficient detail) in the published literature on dividing wall columns.)

To our knowledge, there are no commercial flowsheet simulation packages that offer a DWC as a standard model (rate-based or otherwise). (Models developed in ACM do not meet this standard.) This paper describes just such a model, which, via the CAPE-OPEN framework, can be used in any CAPE-OPEN compliant process flowsheet simulation package. We introduce a new parallel column model in ChemSep that is rate-based, equation oriented and can be used to model DWCs of arbitrary configuration.

We illustrate the model by showing the differences in design that can arise when using these two types of column model. Any DWC with an interior wall (meaning that the wall does not extend all the way to the top and bottom of the column) must have at least one side draw for the intermediate boiling product(s). The optimum location of side draws is one area of column design where a rate-based model should be superior to model based on equilibrium stages.

The model in ChemSep allows for heat transfer across the wall and we compare the predictions of the model to experimental data from the University of Texas at Austin.

**References**

1. Dejanović, I., Matijašević, L. & Olujić, Ž. Dividing wall column—a breakthrough towards sustainable distilling. *Eng. Process. Process Intensif.* **49,** 5pp 59–580, 2010
2. Yildirim, Ö, Kiss, A.A., Kenig, E.Y., Dividing wall columns in chemical process industry: A review on current activities, *Separation and Purification Technology*, **80**, pp 403-417, 2011
3. Kaibel, B. Dividing-Wall Columns, in *Distillation: Equipment and Processes* pp 183–199, Academic Press, 2014.
4. Mueller, I. & Kenig, E. Y. Reactive distillation in a dividing wall column: rate-based modeling and simulation. *Industrial & Engineering Chemistry Research*, **46**, pp3709-3719, 2007
5. Mueller, I., Pech, C., Bhatia, D., & Kenig, E. Y. Rate-based analysis of reactive distillation sequences with different degrees of integration. *Chemical Engineering Science*, **62**, 7327-7335, 2007

# 514003 Saving Potentials by Use of Dividing-Wall Columns for Natural Gas Liquid Separations

**Abstract:**

Recovery of natural gas liquid (NGL) typically uses three or four fractionation columns to separate the natural gas feedstocks into different product streams, i.e. methane(C1), ethane(C2), Propane (C3) or Propane/butanes (C3&C4), C4 and heavier hydrocarbons (C5+). The number of distillation columns required for separating N-products shall be N-1. This is because one conventional column can separate only two products. Different from the conventional columns, a dividing-wall column (DWC) can be used to separate three or more products in a single column and could have significant savings (up to 30%) in both capital and energy costs. However, currently no commercial installation of DWC for NGL fractionations has been reported in the literature.

This paper analyzed applicability of DWC for NGL separations related with liquefied natural gas (LNG) processes. Feed compositions, relative volatilities and product specs have significant impact to DWCâ€™s feasibility, constructability and/or savings. DWC appeared not suitable for demethanizer and deethanizer columns, primarily due to high cost of refrigerant utilities for overhead condensers. Nevertheless, when an NGL unit has both depropanizer and debutanizer columns, such as where a propane stream is recovered as product or refrigerant, DWC can have good potential for savings in energy and/or capital costs, depending on the richness of the feed gases. A step-by-step flowchart has been introduced for determination of DWCâ€™s applicability. Some considerations for DWCâ€™s operation pressures, hydraulic balances, column internals, reflux splitters and control strategies are also discussed.

## 514037 Tower Entry Safety Topics

**Abstract:**

This presentation is about safety courses/training that should/could be provided to people before they enter confined spaces. The audience is assumed to be typically distillation engineers and probably a number of them have not been in a confined space before. The author will talk about the History of Safety Councils, what Training Topics are available at both Floridaâ€™s as well as other Gulf Coast Safety Councils and also cover why many of them have reciprocity.

There will be a mention of such training topics as the importance as Lock out/ Tag out, harness use, air monitoring, hole watch, vessel walk down and fall protection. It will be noted when people need to get this training (how much in advance of vessel entry) and how often they need to be â€œretrainedâ€ so that safety is ever present on their mind during the work day.

# 514038 Practical Tower Safety from the Engineer's Perspective

**Abstract:**

This session will deal with the practical aspects of column entry safety. It is assumed that training has been completed. Topics will include what PPE should be worn and the quality of PPE. This topic will be presented from an engineers perspective.

Topics should include;

PPE

Gas Monitoring and air quality

Respirators

Harnesses, extraction and Life Lines

Climbing inside towers and fall protection

Ear Protection and noises from air movers and air guns

Common hazards (sharp parts, hoses, NORM, knee pads etc)

Inspection - without slowing down the work

## 514039 Practical Tower Safety from the Installer's Perspective

**Abstract:**

This session will present tower entry safety from an Installer's perspective. Topics should cover the following topics and more as well as provide personal experiences.

PPE

Lockout/Tagout

Hole watch

Blinding & double block & bleed & Unit â€œWalk downâ€

Fire Protection

Hydration, Heat Stress, and stamina

Intrinsically safe (what does that really mean)

Claustrophobia and darkness

Asbestos, dust, mold, dirt, oil, sludge etc.

Hot work (sparks, welding bright lights, burns etc)

Safe Work Permits and the permitting process

# 516714 Non-Ideal Two-Phase Flow on Large Diameter Trays: The Stagnant Regions Model versus FRI Experiment

**Abstract:**

Forty six years ago, the Stagnant Regions Model (SRM) was published by Porter, Lockett and Lim in which the impact of severe flow maldistributions on tray efficiency for large diameter trays was investigated. The SRM involves two zones to characterize the liquid flow pattern on a single-pass tray: an active liquid channeling zone between the downcomers and retrograde/stagnant zones at the sides of the tray. It was found that for large diameter sieve trays, a loss in tray efficiency was reported due to liquid channeling and associated stagnant zones. The SRM theory was supported experimentally years later in the Aston University 1.22 and 2.44 m air-water simulator columns (Porter et. al., 1987, 1992 and 1995). At the same time the SRM was published, FRI studies on maldistributed liquid flows on 1-pass sieve trays using hydrocarbon systems in the 2.44 m FRI Low Pressure column showed no loss in tray efficiency both with and without flow control mechanisms (Bell, 1972; Yanagi and Scott, 1973).

The purpose of this paper is to discuss tray geometry parameters employed in the SRM to explain why the model predicted no loss in tray efficiency on the FRI 2.44 m diameter 1-pass sieve tray. One important parameter to be investigated is Width-of-the-Mixing-Zone (WMZ) at the sides of the tray. Arguments will be supported with FRI film footage of liquid distribution both without and with flow control mechanisms.

## 512752 Evolution of Random Packing Performance with a Focus on Geometry

**Abstract:**

The geometries of random packings have evolved significantly over the years to offer a broader range of hydraulic and mass-transfer performance. Packings were originally made from local materials, but new generation packings with engineered geometries have been designed and manufactured since the 1950â€™s to significantly increase specific surface area with improved pressure drop. This presentation will investigate how packing geometry affects form drag and friction drag with the evolution of packings over time, and how they are addressed with fundamental correlations.

Key Words: Random packing, Packed Columns, Pressure Drop, Correlations

*Corresponding author. E-mail: ken.mccarley@fri.org , Phone: 1-405-385-0354, Fax: 1-405-385-0357

# 512806 Computational Fluid Dynamic (CFD) Simulations of Multiphase Flow through Representative Elementary Units (REU) of Structured Packing

**Abstract:**

**Computational Fluid Dynamic (CFD) Simulations of Multiphase Flow Through Representative Elementary Units (REU) of Structured Packing**


Trien (Mikey) Phan and R. Bruce Eldridge

James R. Fair Process Science and Technology Center

The University of Texas at Austin


Current model predictions of the hydraulic behavior and separation performance of structured packings rely on semi-empirical correlations whose ranges of viability are limited to the packings and operating conditions used during correlation development. A more fundamental understanding of the momentum transfer inside structured packings is crucial for improving model predictions of the hydraulic properties of structured packing, which in turn can lead to optimized designs of mass transfer column internals. To this end, we utilized Computational Fluid Dynamics (CFD) to simulate gas-liquid flows inside simplified Representative Elementary Units (REU) of structured packing. Using Siemensâ€™ STAR-CCM+ CFD software, we investigated the impact of packing geometry, such as corrugation angle and flow channel dimensions, as well as fluid properties, such as surface tension and contact angle, on the hydraulic performance of an REU. The modeling results yielded an optimum geometry for the hydraulic performance of structured packing.

# 513259 Evidence for the Need of Probability Density Function Representations of Liquid-Side Mass Transfer Rate Coefficients for Absorption Modeling in Structured Packings


**Abstract:**

Gas-liquid mass transfer parameters of structured packings are generally determined from global absorption or desorption measurements over some bed height, sometimes with treatment of entrance and outlet effects. Correlations of these global parameters are then developed relative to the overall flow conditions and fluid properties and often take the classical form

Eq. 1

where and stands either for the gas phase, G, or the liquid phase, L. These equations are subsequently applied in models for sizing large absorption or distillation columns using local average flow conditions and fluid properties. Rejl et al. [1] have recently shown that the dependencies to the flow conditions in such correlations, while allowing to match global features, such as HETP, do not accurately reproduce axial concentration profiles. They propose a so-called "profile method" to obtain better functions relative to flow conditions. In this work we present other experimental evidence that indicates that correlations of mass transfer parameters derived from a global perspective may be insufficient to properly describe the details of the processes required for column sizing.

We consider the case of physical absorption or desorption. Based on the double-film model the global overall gas-side mass transfer resistance ($1/K_Ga$) is mathematically the sum of the global gas-side ($1/k_Ga$) and liquid –side ($m/k_La$) resistances:

$1/K_Ga=1/k_Ga+m/k_La$ or $R_{G,ov} = R_G+mR_L$ Eq. 2

where $m$ is the slope of the equilibrium curve. This equation is valid if all liquid-gas interfacial areas are subjected to the same mass transfer rates. However, in many situations such is not the case. Haroun et al. [2] have done CFD calculations of film flow over a packing structure and indicate that, in coherence with the Higbie contact time surface renewal model, the liquid side mass transfer coefficient decreases inversely with the distance or time from the renewal point. Recent Xray tomographic imaging [3, 4] and CFD simulations [5,6] indicate the existence of flow heterogeneities leading to rivulets formation and unwetted (dry) regions, even with hydrocarbon liquids that have better wetting characteristics than water.

Now we further consider a case where the global or average $mR_L$ is secondary to $R_G$ . If there exists a local region in the packing where the liquid film velocity is low then it will likely be characterized by a longer renewal time relative to the mean. This will lead to a locally higher $R_L$ that will tend to reduce the overall $K_G$ and hence the mass transfer rate. In contrast, if there is a region with very low $R_L$, say along a rivulet, then there will be no impact on the overall mass transfer rate since the liquid side resistance is already globally low. The opposite will hold for a case where global liquid side resistance is large. Therefore local flow heterogeneities will, under certain conditions, alter the true mass transfer rate relative to one predicted by liquid and gas-side mass transfer parameters determined independently and on a global basis.

To illustrate this fact, we present experiments that have been performed with two different chemical systems to measure the global $k_Ga$ of Mellapak 250.X packing in a 146 mm diameter column. One system is the fast reactive absorption of $SO_2$ diluted with air into a 2N solution of NaOH. This system is well known to be only gas side limited [9]. The second system is the physical absorption of ethanol or methanol diluted with air into water for which the contribution of liquid side resistance to the overall resistance is often small. So from a global perspective, with the reasonable assumption of negligible liquid side resistance, both systems should give the same relation for $Sh_G$ as given in Eq. 1. Yet the values of $Sh_G$ determined from $SO_2$ absorption are 50 % higher than from ethanol absorption (Figure 1). Based on global mass transfer parameters from the literature [7] , under our flow conditions, the contribution of liquid side resistance to the total does not exceed 10% at the highest gas flowrate. The 50% difference is well beyond the experimental uncertainties of the global $k_L$ measurements. For example, it would be necessary to excessively reduce $k_L$ by 4 to 5 times in order to make the data coherent. In order to help understand the different results between the two systems, further experiments were carried out in a 19 mm ID column over a wire gauze high effective area packing (Sulzer EX) at very low liquid loads. The liquid side resistance was measured independently and found to contribute by at most 5% to the overall resistance. In this situation the two chemical systems gave the same $Sh_G$ values within 10%.

Figure 1. Comparison of $Sh_G$ values obtained with different systems over Mellapak 250.X packing at a liquid load of 30 m/h. The values have been normalized relative to the one obtained with $SO_2$ absorption at $Re^{2/3}Sc^{1/3}=100$.

To explain these contrasting results we propose to consider the nature of the liquid film flow. In the wire gauze packing the liquid film spreads very uniformly by capillary action and likely stays very smooth. In the Mellapak 250.X packing the heterogeneities in the film flow are likely to lead to some form of distribution of local $k_L$ or $R_L$ values. The existence of $k_L$ distributions has been discussed in the past by Danckwerts [8] in terms of a distribution of renewal times and how it impacts the expression for the global $k_L$. However no interaction of this distribution with the overall resistance to mass transfer was discussed. If the distribution is described by a probability density function, $f(k_L)$, then the fraction of effective area that is subjected to liquid side mass transfer coefficients of values between $k_L \pm dk_L/2$ is given by $f(k_L)\cdot dk_L$. The average liquid side mass transfer coefficient and overall resistance are then calculated as

Eq. 3

Eq. 4

The distribution function should yield $= k_L$ and, based on the experimental results, $/R_G = 2/3$. As an exercise, we found that a log-normal probability density function on $k_L$ was able to satisfy these conditions whereas simple normal distribution function could not. The true nature of the distribution should be the topic of future work.

In conclusion the local liquid flow heterogeneities that can exist in random and structured packings possibly produce a distribution of local liquid–side resistances. The use of a global liquid-side mass transfer coefficients may then lead to erroneous sizing of physical absorption or distillation columns in situations where both liquid and gas-side resistances contribute to the overall mass transfer resistance. Under reactive absorption the $k_L$ distribution will furthermore affect the local Hatta number ($Ha$) and lead to a distribution of mass transfer regimes. Evaluation of the consequences these distributions should be the topic of future work. Last we conclude that physical absorption methods are not suited for measuring the global $k_Ga$ in random or structured packing when liquid flow heterogeneities exist, instead, reactive absorption of $SO_2$ in NaOH should be used.


[1] F. J. Rejl, L. Valenz, and V. Linek, "Profile Method" for the Measurement of kLa and kVa in Distillation Columns. Validation of Rate-Based Distillation Models Using Concentration Profiles Measured along the Column , *Ind. Eng. Chem. Res.*, 49 (9) (2010) 4383–4398

[2] Haroun, Y., Raynal, L., Legendre, D., Mass transfer and liquid hold-up determination in structured packing by CFD. *Chem. Eng. Sci.*, 75 (2012) 342–348.

[3] S. Schug and W. Arlt, Imaging of Fluid Dynamics in a Structured Packing Using X-ray Computed

Tomography, *Chemical Engineering and Technology* 39 (2016) 1561–1569.

[4] A. Janzen, J.Steube, S.Aferka, E.Y.Kenig, M.Crine, P.Marchot, D.Toye, Investigation of liquid flow morphology inside a structured packing using X-ray tomography. *Chemical Engineering Science* 102 (2013) 451–460

[5] Y. Haroun, L. Raynal, P. Alix, Prediction of effective area and liquid hold-up in structured packings by CFD, *Chemical Engineering Research and Design,* 92 (2014) 2247–2254.

[6] Bing Dong, X.G. Yuan , K.T. Yu, Determination of liquid mass-transfer coefficients for the absorption of $CO_2$ in alkaline aqueous solutions in structured packing using numerical simulations , *Chemical Engineering Research and Design* 124 ( 2017 ) 238–251.

[7] C. Wang, D. Song, F. A. Seibert, G. T. Rochelle, Dimensionless Models for Predicting the Effective Area, Liquid-Film, I*nd. Eng. Chem. Res.*, 55 (2016) 5373−5384. .

[8] Danckwerts, P.V., 1970. *Gas-Liquid Reactions*. McGraw-Hill, New York.

[9] L. Hegely, J. Roesler, P. Alix, D. Rouzineau, M. Meyer, Absorption methods for the determination of mass transfer parameters of packing internals: A literature review, *AIChE Journal* 63 - 9 (2017) 3246–3275.

# Distillation Control – Practical Aspects

**Sigifredo (Siegfried) Nino, P. Eng.**

**Summa Control Solutions Inc.**

**Montreal, QC, CANADA**

**SNino@SummaControlSolutions.com**

**Presented at the 2018 AIChE Spring Meeting, Advances in Distillation I, Orlando, Florida, April 22nd – 26th, 2018**

# Distillation Control – Practical Aspects

Siegfried Nino, P.E. - SNino@SummaControlSolutions.com

Process Control Consultant, Summa Control Solutions Inc., Montreal, Canada

One of the most important aspects of implementing, commissioning or troubleshooting any distillation control strategy, resides in the recognition that modifications to existing equipment are improbable. This paper illustrates several cases where changes to original control strategy designs were made to be able to complete a commissioning or resolve long-standing issues relating to distillation control.

Process control makes a significant contribution of to the improvement of plant operation stability. Improving a unit's stability may appear, at first, to have no benefits to the operation or to the optimization of a chemical processing facility. Yet stability enhances the clarity in the landscape of the process' variables and as a result, opportunities for improvement emerge as oscillations are tamed. Included in this writing are relevant and practical tips to illustrate the benefits of this practice.

## BENZENE TOWER - TEMPERATURE DIFFERENCE VS. COMPOSITION CONCAVE RELATIONSHIP

### Background

This tower is part of a Benzene-Toluene-Xylene (BTX) train at a refinery. Its purpose is to separate the benzene from the toluene. Both products must remain at high purity. The column has 54 trays counted from the top. The feed enters at tray 31 and the benzene product is withdrawn as sidestream from tray 6. The lighter material is removed from the overhead condenser and steam is used to provide the heat input while air coolers are used to condense the overhead vapour. Most of the feed leaves through the bottom product with 0.05% BZ impurity in the Toluene stream specification.

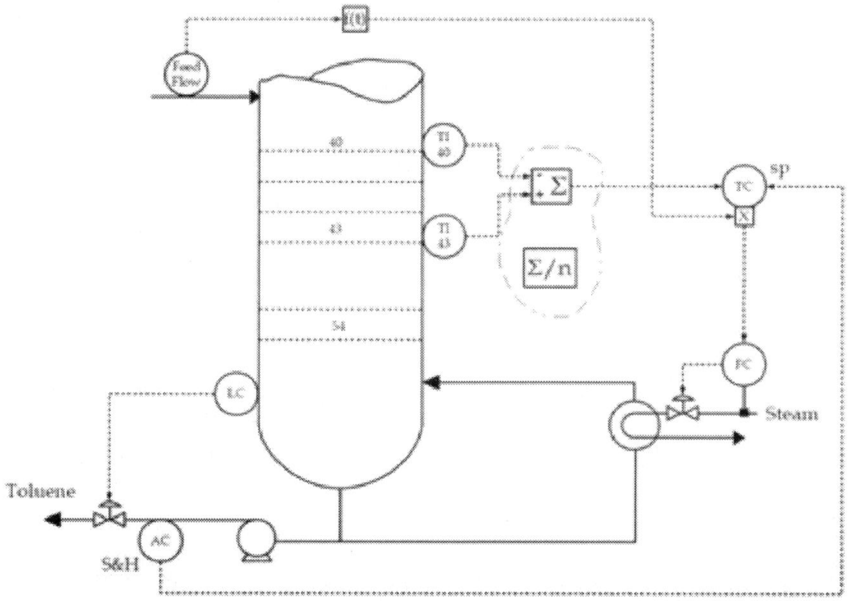


*Figure 1 - Benzene Tower - Bottoms detail*

Two differential temperature measurements were originally used for double-ended composition material-balance control. The top differential temperature (ΔT), is calculated between plates #2 and #14 as the surrogate of toluene impurities in the benzene product. The dual-end composition control of this high-purity distillation column was completed using a bottom section ΔT, calculated by using sensors located three trays apart (Tray #40 and Tray #43). These temperature sensors were RTD (Pt 100 DIN 43760-1968) ranged 0-1600 °F, with an accuracy no better than ±0.25% of span or ±4 °F, for a differential temperature in the range of 4°F, causing a sizeable degree of uncertainty in this measurement.

Figure 2 shows the test performed to qualify and quantify the use of the differential temperature as benzene in the bottom product surrogate. The bottom composition control implemented is the double cascade array shown in Figure 1, where the outmost PID controller operates in a Sample-And-Hold mode synchronized with the "new sample" signal from the bottom product in-line analyzer, setting the setpoint to the differential temperature controller. In turn, this controller sets the steam supply required to keep the tower boilup. In normal operation, a fall in differential temperature is due to the reduction of the light key in the bottom, so the controller reduces boilup, and then both differential temperature and bottom composition will rise to the desired values.

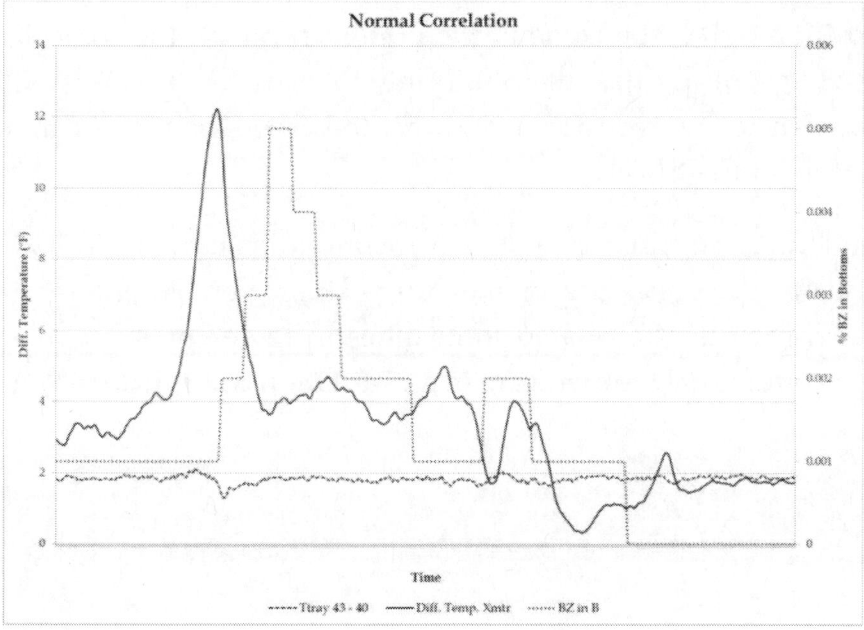


*Figure 2- Benzene Tower - Differential Temperature Test*

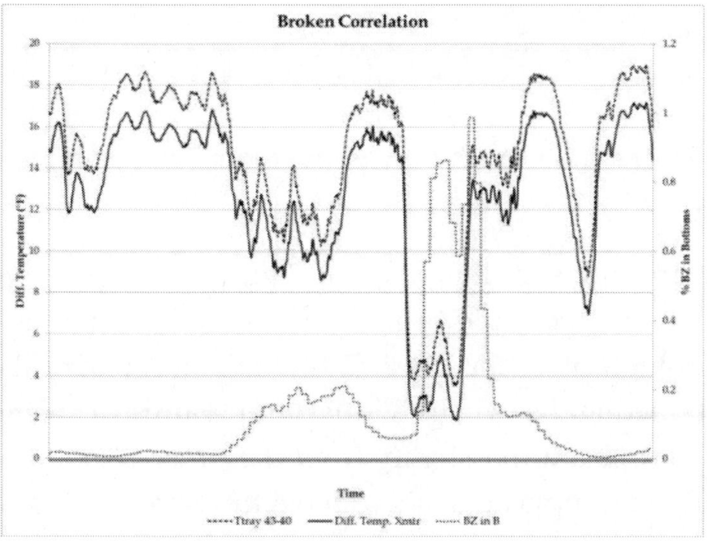


*Figure 3 - Benzene Tower - Breaking ΔT and Composition relationship*

Using a similar analysis designed by Webber (1959), who used the McCabe-Thiele diagram, it was found in the exercise that two orders of magnitude difference in product composition, will cause the Tray40-Tray43 versus Bottom impurities to follow the concave curve in Figure 4. The operation on the left-hand side of the curve means that if the ΔT setpoint equals 10°F and the measurement is 12°F, the controller will need to increase the steam in order to reduce the temperature difference. However, the ΔT could also be at 12°F for a composition of 0.17 Benzene in the bottom product, the controller will then try to reduce the ΔT by cutting the steam to bring ΔT back to 10°F. This will, however, cause the composition to be even farther away from the setpoint, as more impurities will reach the bottom of the tower.

For the sake of completing the topic, it is worth mentioning that Kister (1990) suggested yet another cause for the concavity of the composition versus Differential temperature. He stated that it can also be related to changes in tower pressure, even though it has been said that a differential temperature in a distillation column is immune to pressure variations.

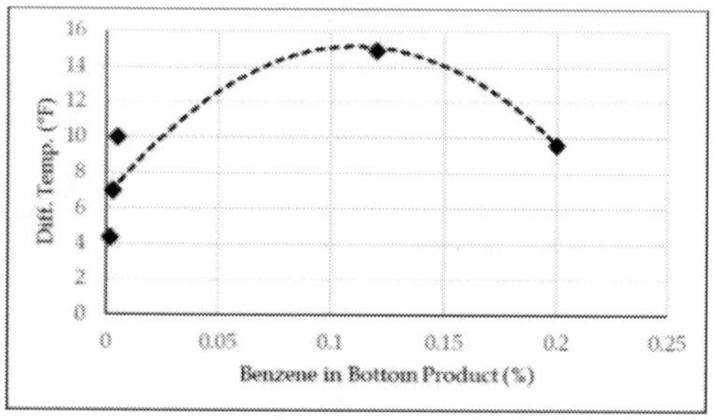


*Figure 4 - Composition vs. dT concavity*

**Solution**

The solution to the inversion in the relationship was overcome by using the temperature sensor's average, as indicated by the cloud in Figure 1.

A relevant issue related to the application is illustrated in Figure 4. Note that at very low composition the temperature difference could be in the order of 4°F. Although the original RTD's where replaced with high precision sensors, the measurement of the differential was done in the DCS by subtracting the two individual temperatures, causing the ΔT to become negative at times, a physical impossibility from the process' point of view. The recommendation for this type of applications is to have the two sensors go into a temperature difference measuring device instead.

The direct benefit of this material-balance dual-composition control in a high-purity distillation application was a 35% reduction in energy consumption.

## TOLUENE TOWER - THE IMPOSSIBLE CONTROL TRAY TEMPERATURE

### Background

This toluene tower is downstream of the benzene tower in a BTX train at a Petrochemical facility. These two towers are thermally integrated; the toluene tower's overhead vapour latent heat supplies the energy input to the benzene tower. Dual-composition control was implemented using Structured Advanced Regulatory Controls.

The tower has 72 trays counting from the bottom. It operates with a flooded reflux drum, most of the product leaves through the top product, and the bottom product is mostly xylene with 1.1% Toluene in the bottom product specification.

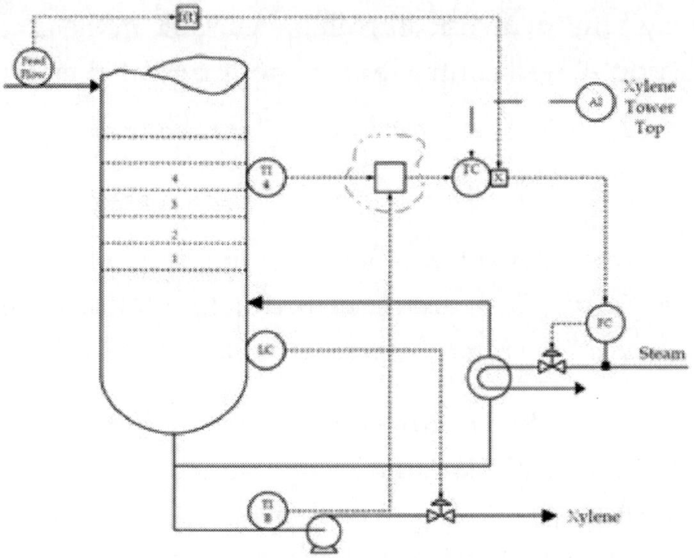


*Figure 5 - Toluene Tower - Bottom Composition Control*

An attempt to control the toluene impurities in the stream towards the xylene tower had been made by using a model-based composition controller that used a Near-Infrared (NIR) analyzer to reset the toluene tower tray #4 temperature surrogate, which in turn manipulated the Boilup to Feed ratio, as shown in Figure 5.

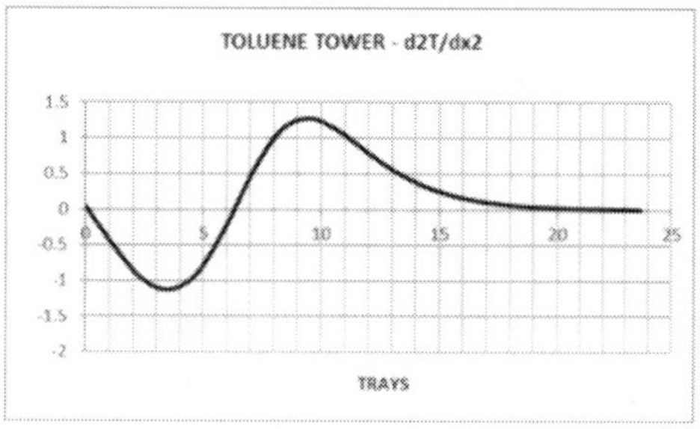


*Figure 6 - Toluene Tower - Second Derivative of Temp. Profile*

The controller didn't really work. In steady state operation, it would work fine, but every time there was an enthalpy disturbance, the composition would follow such extreme excursions that it would take operators several hours, typically a whole shift, to return the composition to specification. This problem didn't stem from the use of a Model-Based controller. It came from the whims of an uncontrollable variable in the inner loop of the cascade. This is to say that Model-Based controllers are not magical; they will work fine when they are placed in the right position in the control hierarchy.

**Solution**

There were two causes in this problem: Control interaction in the benzene tower, and control plate temperature hypersensitivity. Both are described below followed by the implemented solution.

The paper that presented this case (Nino, 2011) showed that by using the relative gain analysis, the benzene tower's original conventional dual composition configuration control LV (Reflux set to control top impurities and Boilup to control bottom impurities), resulted in severe interaction causing great disturbances that were propagated to the toluene column through the heat integration system. As detailed in the aforementioned paper, this problem was successfully resolved by implementing material-balance dual-composition control in both towers.

The investigation into the second source of the problem started by taking the tower temperature profile, available from a tray-to-tray simulation, to confirm that the temperature sensor was properly placed in the tower. Figure 6 shows the plot of the second derivative of the tower temperature profile and confirms that the location of the temperature sensor at tray #4 corresponds to a point of maximum sensitivity in the tray-to-tray compositions changes. In the actual application, however, all the attempts to control this tray temperature were unsuccessful. The clarification to this riddle came from the plotting of Tray Temperature versus bottoms product Impurity, illustrated in

Figure 7. It became clear that there are hypersensitivity and even relation reversal, that lead to conditional stability of the controls.

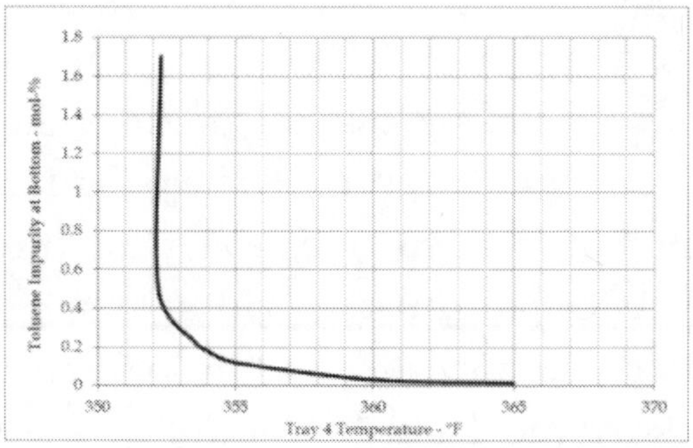


*Figure 7 - Toluene Tower - Bottom Temperature vs. Composition*

An alternate temperature surrogate was found in the bottom product temperature, indicated as TI-B in Figure 5. Testing proved that there is a repeatable relationship with the desired composition. The controller was commissioned for the new configuration and it has been working for more than half a decade.

As mentioned before, a NIR analyzer sampling the Xylene tower's top product was being used for resetting the temperature surrogate. However, this would introduce unwanted disturbances to the control, due to the low analyzer's signal-to-noise ratio. The facility then chose to have the setpoint done by the operator and based on daily lab testing.

The application of this strategy resulted in over $1 MM savings in energy usage and reduction of product recycle. See Nino (2011) for details.

## DEISOBUTANIZER – TRIPLE SPLIT RANGE

### Background

This sidestream alkylation Deisobutanizer (DIB) operates at a recently commissioned oil refinery. The issues: The original overhead accumulator level control was done through the sequencing of three valves, and the technology of the flowmeters selected for the application make them unreliable for this application. This is a work in progress as a material balance control will be needed to optimize the tower operation.

The overhead accumulator is the liquid storage supply for the reflux and the column top product, isobutane in this case. There are two different destinations for the isobutane, alkylation plant and storage. The first one being the preferred route, and the second the excess production being sent to tanks. The drum level control strategy was designed and configured to be a triple split-range scheme: 0-33%, 33-66%, and 66-100%. The first part of the split was supposed to keep a constant flow to the alky plant once the controller output exceeded 33%, and the remaining 66% controller output should have taken care of any variability in the reflux drum inventory.

The commissioning of the DIB was done before that of the alky plant, unveiling the inability of the strategy to keep the overhead drum under control. The operators ended up operating the loop in manual, which implied frequent moves intended to prevent the drum from flooding or emptying. The abnormal situation did not change even after the start-up of the Alky Plant. The strategy could not work and the operators were continuously changing the controller output to keep the drum "in control".

**Solution**

In a process where there is a pump attached to a storage, the average level responds as an integrator. Thus, the process becomes non-self-regulated and in the absence of an able controller, it will require constant operator attention.

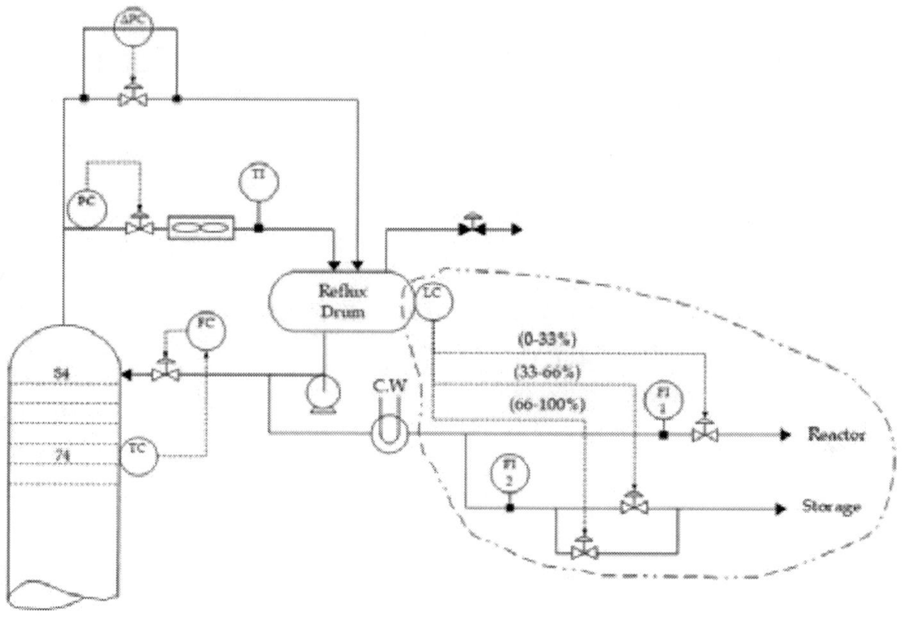


*Figure 8 – Alkylation Deisobutanizer*

The problem was resolved by assigning the first control valve to the stand-alone flow control loop, FC-1 in Figure 9. This ensures a constant feed flow to the alky plant. The other two valves, being the same size and having linear trim were assigned to be the manipulated variable for the level control loop, as shown in Figure 9. A second control loop would be the right solution to implement, but the operation of the second flowmeter

(FI-2 in Figure 9) is not reliable because "Vortex flowmeters require turbulent flow to operate and will cease to read as the fluid transitions from the turbulent flow regime to the transitional or laminar flow regime. This is called the "low-flow cut-off" point of the meter, and flow rate measurement below this point is not possible" (Mc. Millan et al,

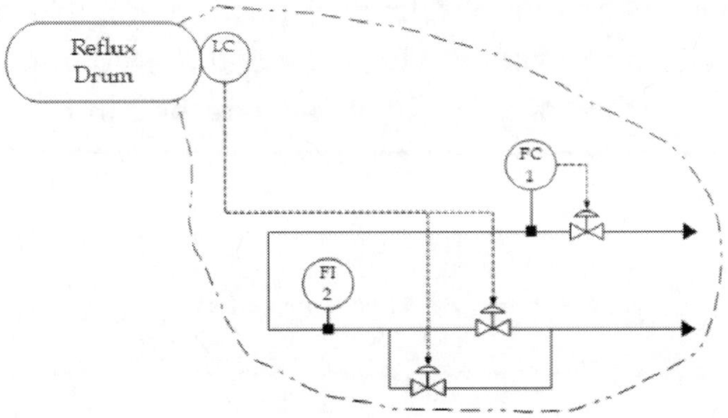


*Figure 9 - Deisobutanizer - Reflux drum level control*

2013). Whilst the flow rate to the alky plant is maintained over the minimum "readable", that is not always the case for the flow to storage tanks.

As mentioned above, a better solution could be implemented by using a material balance control to coordinate the flow loops, which would require reliable flowmeter operation.

Again, a lesson that should have been learnt a long time ago: The open-loop unstable systems are operationally critical, and although operators can take care of them in exceptional circumstances, when it becomes the norm, it's bound to cause more noticeable disturbances to the unit or the plant operation.

The benefits in this case include reducing operator attention and time spent in this unit, freeing the Operator to maintain awareness of all units assigned to him/her.

## OSCILLATING ATMOSPHERIC TOWER

### Background

The crude unit at this refinery was subject to 50-60 minutes oscillation (Figure 11) thought to be originated in the fuel gas network. The oscillation was traced back to the crude tower. It was found out that in the crude unit original control strategies design, the tower pressure control, indicated with a hash background circle in Figure 10, was associated to a manipulated variable that had virtually no effect on the controlled variable.

A validation test performed in the unit proved that the chosen pressure controller was more related to maintaining a constant compressor suction pressure and that its effect on the crude tower pressure was insignificant.

**Solution**

Distillation column pressure is one of the most difficult distillation variables to control. The pressure in a distillation tower can be varied with Temperature, Volume or Mass, the first two are not really available so the only option available is the control the vapor inventory.

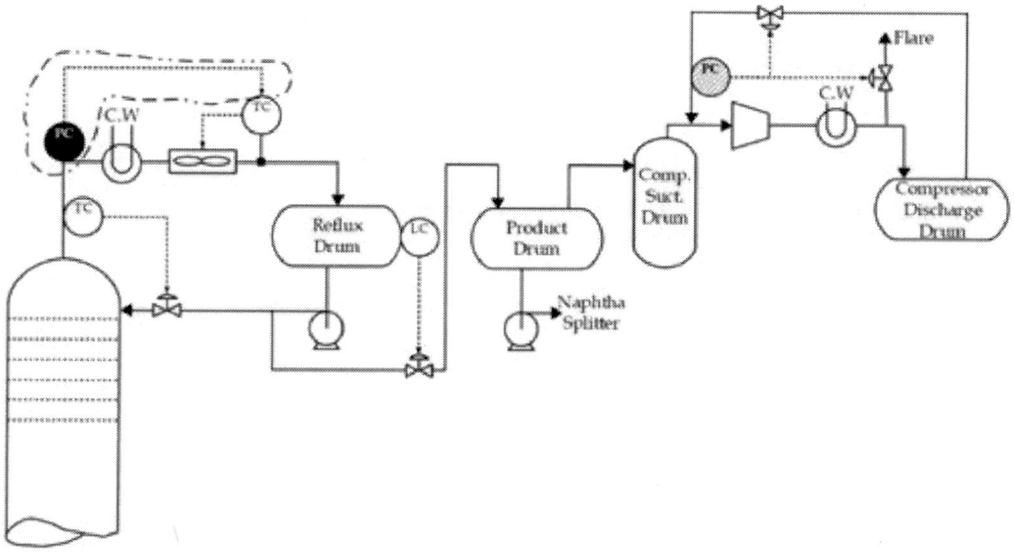


*Figure 10 - Crude tower - Tower Pressure Control- (C.W. is Cooling Water)*

The condenser cooling gave an additional degree of freedom to the system, and the relevant testing showed that the tower pressure would be successfully controlled by manipulating the Air Coolers' exit temperature in a cascade control with the pressure controller, as emphasized by the black background circle, on the left-hand side of Figure 10.

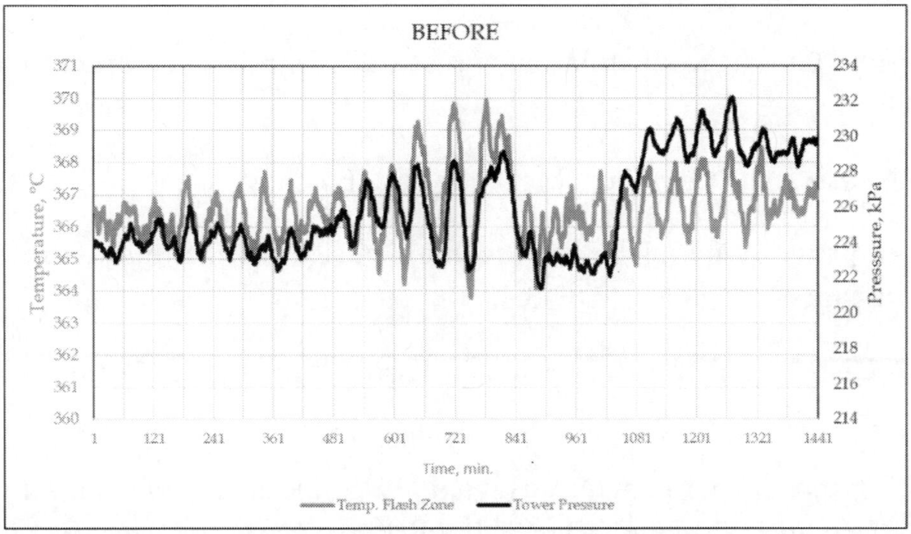


*Figure 11- Crude Tower Oscillating Pressure – Before*

After the commissioning of the new pressure controller, the entire tower temperature profile dramatically improved (Figure 12). It was much easier to separate the different cuts as the cuts overlapping was reduced.

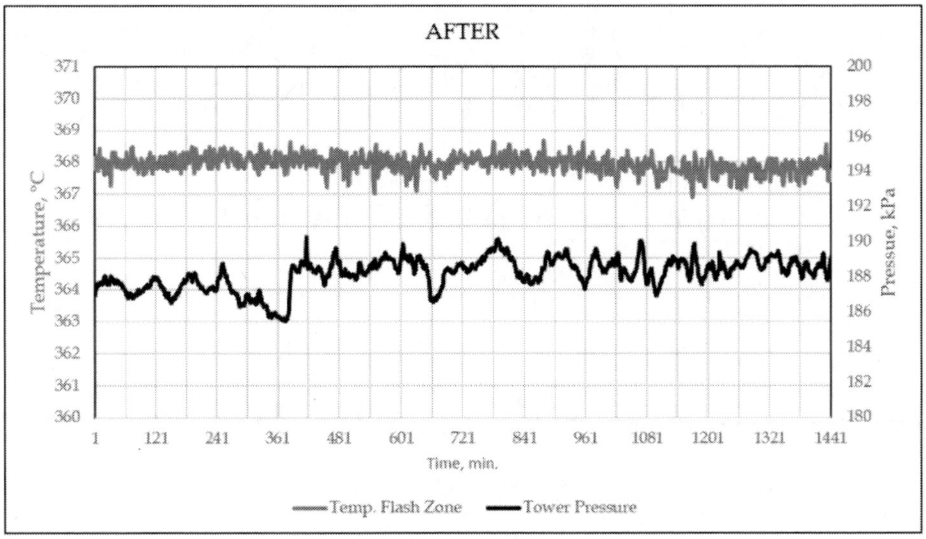


*Figure 12 - Crude Tower Oscillating Pressure – After*

This resulted in the pressure standard deviation reducing from 8 kPa to 0.5 kPa, i.e. sixteen (16) times reduction, and the peak-to-peak deviation was reduced by one order of magnitude, from 34 kPa to 3.4 kPa.

## BIBLIOGRAPHY AND REFERENCES


[1] Chin, T. G. (1979) "Guide to distillation pressure controls", Hydrocarbon Processing, October

[2] Kister, H. Z. (1990) *Distillation Operation*, McGraw-Hill, Boston

[3] Nino, S. (2011) "Improve material balance in high-purity distillation control", Hydrocarbon Processing, November, p.83-88

[4] Nino, S. (2017) "Distillation Control: Still an Issue", in Kister Distillation Symposium 2017 proceedings

[5] Rademaker, O.; J. E. Rijnsdorp and A. Maarleveld (1975) *Dynamics and Control of Continuous Distillation Units*. Elsevier Scientific Publishing Company, Amsterdam. P. 375

[6] Shinskey, F. G. (1984) *Distillation Control: For Productivity and Energy Conservation*, 2nd Edition, McGraw-Hill

[7] Sloley, A. W. (2001) "Effectively Control Column Pressure", Chemical Engineering Progress, January

[8] Webber W. O. (1959) "Control by temperature difference?" Petroleum Refiner V.38, No.5

[9] McMillan G. K.; Vegas H. (2013) 101 Tips for a Successful Automation Career, ISA

## 514036 Packed Tower Separation Efficiency Problem Diagnosed Using Gamma Scanning Technology

**Abstract:**

Dow Chemical was experiencing separation efficiency problems in one of its product towers. Tower internals consisted of four beds of structured sheet metal packing. Process calculations showed the tower was operating with 40% less equilibrium stages than design. Liquid maldistribution was expected and gamma scans were completed. In addition to using the standard four path grid scan layout, two additional scan lines crossing through the center of the tower were done to increase the chances of detecting maldistribution.

Effort was made to quantify the liquid maldistribution based on the gamma scan results to serve as input into a maldistribution sensitivity study. The conclusion from this analysis was the estimated maldistribution was great enough to have a significant effect on the tower efficiency. Not able to discern the root cause of the liquid maldistribution at the time (e.g. distributor design, fouled distributors, installation problem, problem in the packing, etc.), new high quality liquid distributors were designed and installed.

After installation of new liquid distributors, the tower started up and the separation efficiency improved to 98% of the design stages. This presentation includes discussion and results focused on the gamma scans and what can be done with gamma scan data to help with troubleshooting liquid maldistribution.

# 510574 UFM ™ Trays Expand into the Fixed Valve Arena with the UFM ™ AF

**Abstract:**

The UFM tray has been accepted globally as the next generation movable valve with over 350 towers in operation globally. It is only logical that this outstanding device be further developed for the Fixed Valve Arena. The fixed lift UFM is now available for those applications that have a tendency to foul yet still need the high performance and high capacity that is provided by the UFM valve. This paper will show the capacity, performance and pressure drop of the new, fixed anti-fouling UFM™ AF valve.

# 512680 Deep-Seal Sieve Trays for Improved Scrubbing Performance

**Abstract:**

**Deep-Seal Sieve Trays for Improved Scrubbing Performance**

In trayed columns, outlet weirs on the order of 2" high are commonly used to maintain liquid level on the deck. Shorter weirs, at least in froth regime operation, can reduce efficiency, and higher weirs are generally thought to increase pressure drop with little benefit in efficiency.

In some applications, however, pressure drop is actually advantageous. For example, packed towers are a common choice for scrubbing of acid gases because of their low pressure drop, but are less effective at scrubbing particulates. As particle diameter decreases, the pressure drop required to get good contact between the particulate and the scrubbing liquid, and therefore efficient removal, increases. For really small droplets, the pressure drop required can be substantial, well beyond what a typical packed column, or even a trayed column, can deliver.

This paper describes an application of deep-seal sieve trays, which have been successful in improving scrubbing performance of a formerly-packed column in an operating facility. It summarizes parameters necessary for successful design and operation of these trays, based on previously-published but not-widely-understood criteria. Because the weir depth, at 12", is outside the range over which typical tray correlations were developed, the fact that these operate well may have additional implications on how some well-established correlations extrapolate.

## 509483 Revamping a Giant C3-Splitter Tower

**Abstract:**

With an abundance of affordable natural gas and continued growth in the utilization of poly-ethylene and poly-propylene several very large splitter towers were built in the recent past. The purpose of these splitters is to produce polymer grade olefins as an overhead product and to minimize the fraction of the olefins in the bottoms product. Koch-Glitsch recently revamped a large diameter C3-splitter to increase the capacity and improve the propylene purity and recovery. This presentation highlights the steps that were taken in this revamp. This includes large scale pilot studies, CFD calculations, innovative designs, and dealing with challenges during the installation process. This revamp was successful and demonstrates that it is possible to achieve high capacity and efficiency in large diameter towers by carefully choosing the fractionation devices.

## 513013 A Design and Computation Software for Separating Multicomponent Mixtures

**Abstract:**

Separation is one of the most frequently used processes for separation of multicomponent mixtures in the chemical industries, including oil and gas, air separation, biochemical, pharmaceutical, etc. Multicomponent mixtures are separated in sequential steps in multiple distillation columns, referred to as distillation configuration. The number of configurations that are feasible for achieving the separation task increases rapidly as the number of components in the feed mixture increases. Also, performing rigorous calculations for validating the design and economics of a distillation configuration is tedious, and there is no systematic way for identifying alternative configuration options without performing exhaustive calculations. For these reasons, it is challenging for practitioners to retrofit the configurations currently in operation or find the most economic configuration(s) among the vast search space of configurations in the early stages of process design.

We propose a short-cut method by combining Underwoodâ€™s method for estimating vapor flows in pinched columns with mass balance equations to estimate the flows and vapor duties in a distillation configuration. This method reduces the computation load drastically while maintaining a good approximation compared to the rigorous calculations. For a feed mixture of $n$ components, the separation task can be achieved in $n$-1 columns, referred to as basic configuration, or in less/greater than $n$-1 columns, referred to as non-basic configurations. The short-cut method gives us the ability to search within the basic and non-basic search space for an optimal configuration that has minimum vapor duty, exergy loss, and/or capital and operating cost. Therefore, a systematic tool that suggests attractive configurations based on recent advances in distillation technology would have a large impact on energy, environment and economic aspects of the U.S. manufacturing sector.

In this work, we are proposing a design and computation software for identifying the attractive basic configurations that can be used by both industrial practitioners and academic peers. The underlying problem is formulated as non-linear programing (NLP) and mixed-integer non-linear programing (MINLP). Our state-of-the-art software solves the underlying problem to global optima using GAMS/BARON, and displays the attractive configurations that address the userâ€™s needs in a friendly user interface. The visual output of the software not only helps in absorbing the results, it allows the practicing engineers to explore various operable and dividing-wall column analogs. The software is based on our recent advances in developing efficient algorithms for identifying implementable energy-efficient low-cost multicomponent distillation trains.

## 512610 New Test Systems for Distillation Efficiency Experiments at High Liquid Viscosities

**Abstract:**

Although several physical properties govern the design and performance of distillation columns, the effect of liquid viscosity is critical as it influences both mass transfer and hydrodynamic aspects of distillation columns. While there has been research to analyze the effect of viscosity on distillation, the majority of the research has been conducted with liquid viscosities less than 1.0 cP. There are a few articles in the literature discussing the effect of liquid viscosity greater than 1.0 cP on sieve tray efficiency using glycerol/water mixtures of varying glycerol concentrations[1]. However, this may not represent a realistic scenario for distillation, as the relative volatility of glycerol/water mixture is well above the typical range of 1.1 to 2.5. Recently, BradtmÃ¶ller and Scholl proposed a new test system â€“ isobutanol/tert-amyl alcohol, to study the effect of liquid viscosity on distillation efficiency from 1.0 to 4.0 cP. Although this is a useful system to study the effect of viscosity, additional test systems are necessary for decoupling the effect of viscosity from relative volatility. Therefore, in this study, we propose two new test systems that are relevant to high viscosity distillation conditions. The test systems were identified based on a comprehensive search and the selection criteria included (a) relative volatility in the range of 1.1 to 2.5 (b)boiling temperatures from 25 to 160°C to enable laboratory scale efficiency experiments at atmospheric and vacuum pressures in glass columns (c) viscosities more than 1.0 cP at these boiling temperatures. The newly selected test systems are cyclopentanol/cyclohexanol and tert-butanol/tert-amyl alcohol.

Vapor-liquid equilibrium data for the new systems were experimentally measured using a newly installed vapor-liquid equilibrium apparatus. A validation of the newly installed apparatus was first conducted prior to performing new experiments. The viscosities of these new test systems were also measured to accurately model the non-linear dependence of viscosity with temperature and composition. The new test systems would enable collecting efficiency data at viscosities ranging from 1.0 to 4.5 cP. These data are crucial in order to evaluate the validity of existing efficiency correlations as well as to develop improved efficiency correlations for highly viscous systems.

[1] Mahiout, S. and A. Vogelpohl, *Mass Transfer in High Viscosity Media.* Chemical Engineering and Processing, 1984. 18: p. 225-232.

# 512403 Reboiled Refinery Sidestrippers Considered As a Special Case of Thermally Coupled Columns

**Abstract:**

In the last five years both academic and industrial engineers have made leaps in understanding and implementing thermally coupled columns. This has the potential to transform distillation in many industries. However, one area where thermally coupled columns has been long established is in the use of reboiled sidestrippers on crude distillation columns.

Often reboiled sidestrippers are not thought of this way, but they are one of the oldest and most established practical applications of thermally coupled columns. One factor limiting their understanding is that refinery services have product specifications set on performance properties of broad boiling-range product streams rather than by purity targets based on a few components.

This analysis examines the following elements of reboiled sidestrippers;

- survey of industry practice on use.
- capacity and yield benefits or costs.
- heat source analysis and practical limits.
- stastical analysis of economic factors favoring them.
- secondary factors in selection and application.

# AUTHOR INDEX